Sustainable Textiles: Production, Processing, Manufacturing & Chemistry

Series Editor

Subramanian Senthilkannan Muthu, Head of Sustainability, SgT and API, Kowloon, Hong Kong

This series aims to address all issues related to sustainability through the lifecycles of textiles from manufacturing to consumer behavior through sustainable disposal. Potential topics include but are not limited to: Environmental Footprints of Textile manufacturing; Environmental Life Cycle Assessment of Textile production; Environmental impact models of Textiles and Clothing Supply Chain; Clothing Supply Chain Sustainability; Carbon, energy and water footprints of textile products and in the clothing manufacturing chain; Functional life and reusability of textile products; Biodegradable textile products and the assessment of biodegradability; Waste management in textile industry; Pollution abatement in textile sector; Recycled textile materials and the evaluation of recycling; Consumer behavior in Sustainable Textiles; Eco-design in Clothing & Apparels; Sustainable polymers & fibers in Textiles; Sustainable waste water treatments in Textile manufacturing; Sustainable Textile Chemicals in Textile manufacturing. Innovative fibres, processes, methods and technologies for Sustainable textiles; Development of sustainable, eco-friendly textile products and processes; Environmental standards for textile industry; Modelling of environmental impacts of textile products; Green Chemistry, clean technology and their applications to textiles and clothing sector; Eco-production of Apparels, Energy and Water Efficient textiles. Sustainable Smart textiles & polymers, Sustainable Nano fibers and Textiles; Sustainable Innovations in Textile Chemistry & Manufacturing; Circular Economy, Advances in Sustainable Textiles Manufacturing; Sustainable Luxury & Craftsmanship; Zero Waste Textiles.

More information about this series at https://link.springer.com/bookseries/16490

Subramanian Senthilkannan Muthu
Editor

Sustainable Approaches in Textiles and Fashion

Manufacturing Processes and Chemicals

 Springer

Editor
Subramanian Senthilkannan Muthu
SgT Group & API
Kowloon, Hong Kong

ISSN 2662-7108 ISSN 2662-7116 (electronic)
Sustainable Textiles: Production, Processing, Manufacturing & Chemistry
ISBN 978-981-19-0540-7 ISBN 978-981-19-0538-4 (eBook)
https://doi.org/10.1007/978-981-19-0538-4

This Springer imprint is published by the registered company Springer Nature Singapore Pte Ltd.
The registered company address is: 152 Beach Road, #21-01/04 Gateway East, Singapore 189721, Singapore

Contents

About the Editor

Dr. Subramanian Senthilkannan Muthu currently works for SgT Group as Head of Sustainability, and is based out of Hong Kong. He earned his Ph.D. from The Hong Kong Polytechnic University, and is a renowned expert in the areas of Environmental Sustainability in Textiles & Clothing Supply Chain, Product Life Cycle Assessment (LCA) and Product Carbon Footprint Assessment (PCF) in various industrial sectors. He has five years of industrial experience in textile manufacturing, research and development and textile testing and over a decade's of experience in life cycle assessment (LCA), carbon and ecological footprints assessment of various consumer products. He has published more than 100 research publications, written numerous book chapters and authored/edited over 100 books in the areas of Carbon Footprint, Recycling, Environmental Assessment and Environmental Sustainability.

Environmental Sustainability Requirements in the Ready-Made Garment Industry

Rezaul Shumon and Shams Rahman

Abstract Environmental sustainability has become a major concern in the fashion industry as it pollutes the natural environment significantly. Due to this concern, big retailers are under intense pressure from various stakeholders such as consumers, governments, non-government organisations, to address the environmental damages caused by their operations. In addition, to comply with regulations imposed by governments, retailers often come up with their environmental requirements and pass them to the next tier of the supply chain through supply chain contracts. This chapter unveils what happens to these requirements when it passes through upstream in the supply chain. Through a case study, it was shown that Bangladesh ready-made garment industry is facing extraordinary changes in environmental requirements and more importantly these requirements become even more stringent when it reaches the suppliers. It was revealed that there are several factors why the requirements become stringent, and this study suggests how Bangladeshi garment manufacturers can handle such stringent environmental requirements.

1 Introduction

This chapter will discuss a specific issue related to environmental sustainability in Bangladesh ready-made garment industry. The big retailers such as H&M, Zara, Adidas and Walmart have sourcing bases in Bangladesh. These brands and retailers are under pressure to maintain social and environmental sustainability standards from government and non-government stakeholders, media, and competitors as well. However, understandably these brands largely depend on their suppliers to maintain their environmental performance standards. Therefore, being a major global RMG supplier country, Bangladesh is under pressure and facing uncertainty as these environmental practices are not familiar to their RMG manufacturers. The pressure is

R. Shumon · S. Rahman (✉)
Department of Supply Chain and Logistics, College of Business and Law, RMIT University,
Melbourne, Australia
e-mail: shams.rahman@rmit.edu.au

© The Author(s), under exclusive license to Springer Nature Singapore Pte Ltd. 2022   1
S. S. Muthu (ed.), *Sustainable Approaches in Textiles and Fashion*,
Sustainable Textiles: Production, Processing, Manufacturing & Chemistry,
https://doi.org/10.1007/978-981-19-0538-4_1

mounting from downstream buyers to upstream suppliers to address rapidly changing environmental requirements due to the dynamic business environment such as regulation changes, rapid innovation, etc. Therefore, the main objective of this chapter is to shed the light on the dynamics of environmental requirements from buyers and their implications in the context of suppliers such as the Bangladesh RMG industry. Specifically, the objective is to understand why the environmental requirements become 'stringent' and what Bangladesh can do in response to these environmental requirements and sustain in business. This chapter comprises the following sections:

- Sustainability in Bangladesh ready-made garment industry
- The stringency of buyer environmental requirements
- Factors of Stringency
- Adopting Stringency
- A case study in Bangladesh ready-made garment industry
- Conclusion.

The next section will give an overview of sustainability practices and initiatives in Bangladesh ready-made garment industry.

2   Sustainability in Bangladesh RMG Industry

Bangladesh is a leading example of low-income countries where RMG is considered as a pillar of the country's economic growth. This industry is accounted for about 13% of the country's GDP which can be regarded as the pillar of Bangladesh's economy [26]. While the cheaper production cost is the main leverage that buyers can take advantage of, however, managing the supply chain in Bangladesh is a major challenge for their sourcing strategy. Operational issues such as quality and timely delivery have been a major concern previously due to political instability and lack of infrastructure. Bangladesh is progressing on operational aspects slowly; however, this country is infamous for sustainability concerns, namely, social and environmental issues.

The sustainability concern was exacerbated by some major industrial accidents in Bangladesh such as Rana Plaza and Tazrin Fashions which have caused thousands of deaths. As a result, buyers are reasonably more conscious of compliance with sustainability when they make their sourcing decision from developing countries like Bangladesh. This often results in incorporating stricter environmental standards in their supplier guidelines. Inevitably, it also reflects on the contract with their suppliers especially if they are from developing or least developed countries. While social sustainability issues such as child labour, workplace safety, minimum wage were always been a concern, however, buyers are also facing intense pressure to address environmental sustainability issues as textile processes discharge hazardous materials which result in a significant threat to health and the environment. Therefore, buyers are putting some standards and, in many cases, their requirements to follow by their suppliers. As Bangladesh is now a popular destination (3rd largest exporter of RMG products) of prominent retailers across the globe, understandably facing buyer's pressure to meet environmental sustainability requirements. Most Bangladeshi companies are looking for some sort of environmental certification to satisfy their customer. For example, Leadership in Energy and Environmental Design (LEED) is a certification by the United States Green Building Council that certifies a company for its efforts to create healthy, liveable and sustainable spaces. This is one of the most contemporary and frequently asked sustainability requirements by retailers. As a result, up until today, 144 companies in Bangladesh are LEED-certified and 39 of them are rated into the highest category (LEED Platinum). In the list of top 10, nine are from Bangladesh [3]. This indicates the quest of how manufacturers in Bangladesh are striving towards meeting environmental requirements from their buyers. A report by McKinsey and Company [22] indicates that Bangladesh will be facing even more pressure from buyers in terms of environmental sustainability issues in the future. This is true for other countries such as Vietnam, Cambodia and Myanmar as well.

3 Stringency of Environmental Requirements

Large firms and corporations such as multinational companies or retail giants naturally face more intense scrutiny from the government, media; NGOs than their suppliers. They are held more accountable for their suppliers' behaviour and practices [4–6]. This is mainly because of their customer-facing position in the supply chain. For example, after the Rana Plaza incident, most of the big brands and retailers were under huge criticism and pressure from customers and other stakeholders. This is true for environmental issues as well in a sense when suppliers do environmentally harmful activities that impact their buyer's reputation in the global market. Therefore, buyers develop their own environmental requirements framework considering all the local and global regulations ask their suppliers accordingly. The requirements are passed to the next tier of the supply chain through supply contracts, and known as 'green supply' in literature [14]. Introducing environmental requirements into supply contracts is relatively a new practice. Nevertheless, governments and stakeholders are compelling manufacturers to rethink their strategy regarding environmental sustainability of their business and introduce environmental requirements in their supply chain as a result of several *incidents* such as once specific pollutant caused the death of more than eight thousand people in China, Brazil's 'worst-ever environmental disaster'[8], and *initiatives,* e.g. Dutch customs stopped 1.3 million PlayStation to be imported due to level of cadmium that exceeds new requirement [5]. Some prominent examples of the responses to these initiatives are, for instance—Ford mandates ISO14001 as part of their supply agreement for all of their suppliers [10], Mitsuhbishi, BMW and Toyota incorporated supplier activities in their statement of environmental responsibilities, Sony's 'Road to Zero'—Global environmental plan is already well regarded [36]. A comprehensive ethical sourcing guideline has been introduced by

Starbucks [37]. Through green supply, companies set targets to meet their corporate social responsibility, increase reputation in the market, reduce unnecessary waste and increase flexibility to respond when new environmental regulations are in effect [21]. The aim of environmental requirements is primarily to green the supplier processes or activities and the ultimate objective is to improve environmental performance. Generally, these environmental requirements fall under the realm of

- Standards
- Targets
- Innovation
- Reduction (emission, etc.)
- Recycle, Remanufacturing.

Nevertheless, in recent years, product-based regulations have been changed significantly to address the concern over sustainability and climate changes [16]. Such regulatory changes to improve environmental performance flow back upstream in the supply chain and create uncertain consequences for different tiers of suppliers [16]. Much of this is because when these changes are reflected on the supply contract as buying specifications, they do not necessarily match the original regulations and protocols [12]. As a result, environmental performance requirements may become more stringent for upstream suppliers [16]. This stringency is mainly caused by indirect regulations from the buyers—a practice that is similar to the 'rationing and shortage gaming' concept of [15].

In addition to this, buyer firms tend to address the regulatory changes well before the deadlines set by the regulators. Some reasons for creating this 'time buffer' [16] are—supplier commitment towards environmental practices is rated as poor [40], suppliers are not willing to accept rather reluctant [13], and limited supplier capabilities, all of which ultimately delays the implementation [17]. Therefore, as risk-averse, buyers tend to shift their original deadline to an earlier date when they pass it to their supplier, thus becoming more stringent [16].

Large firms are under more intense pressure from the stakeholders for their environmental performance in addition to firms that operate in high pollutant industries as they tend to attract media focus [7, 25] (Bansal and Roth 2000; Darnall 2006). Industry bodies often recommend their members to adopt conventional management standards so they can influence the perceptions of regulatory bodies and other stakeholders about their operations. Nevertheless, such management standards do not necessarily produce consistent evidence of environmental performance improvement, often they fail because of the narrow focus of the standards, doesn't fit well to the needs, and only focus on one performance metric [7, 35]. This leads big retailers to develop their operational capabilities beyond compliance to traditional standards; in fact, some firms define their standards [24]. Consequently, in a competitive environment, leader firms set their performance and innovation targets higher than their competitors in the market [29]. However, developing beyond basic capabilities requires higher investment in physical and knowledge resources. Many high-performance targets of large firms may create complexity and ambiguity because their suppliers are usually smaller and technically less capable, more risk-averse, and have limited resources [6]. This situation is supported by the findings of [16] who argued that transferring new buyer environmental requirements to the suppliers may create complex dynamics in the supply chain because of various types of responses from suppliers.

Traditional environmental requirements from buyers such as ISO 14001 standard are well known and easy to understand and communicate [6, 7]. Nonetheless, requirements that are coming from the buyer's own 'regulation' are very specific. Understandably, these requirements are not well established in the market and therefore unknown to the suppliers. As a result, developing knowledge and physical resources aligned to the requirements is critical for suppliers [16, 28, 45]. Redesigning the process could also be required to meet the changed requirements. Suppliers perceive these requirements as complex and more challenging to address due to the lack of technical capabilities, financial and human resources, and risks associated with the investment [6]. It has been argued that suppliers may need a new set of physical capabilities, deploy additional resources, in addition to giving a face-lift in collaboration with buyers for technical and managerial assistance. Collaboration with buyers results in supports often in the form of environmental training, education and technical support.

It is evident from supply chain literature that there is a significant presence of 'stringency'; however, it is unknown why it does happen. In the next section, we will explore why the 'stringency' of environmental requirements happens in a supply chain.

4 Factors of Stringency

There are several reasons why stringency may arise in a supply chain. For example, unanticipated changes in the business environment, regulation changes, implementation complexity to name a few. We will further explore the different factors of stringency in this section.

4.1 Uncertain Changes in Requirements

Despite some common elements, environmental regulations are significantly different across countries and continents. Bangladesh garment industry is exposed to most of the environmental regulations around the world since manufacturers export to most of the countries of the world. Therefore, garment suppliers in Bangladesh need to comply with many different regulations and face a variety of environmental requirements from their buyers. In many circumstances, it is uncertain for suppliers what are environmental requirements imminent to them. For example, buyers frequently ask suppliers to reuse their wastewater for which they have little or no preparation at all. Suppliers often are not familiar with this kind of requirement before, are not informed well before it needs to be addressed, and further do not know even what changes are going to happen in the requirement next year. This unpredictability leads to environmental uncertainty in suppliers' operations. In addition, frequent changes in environmental requirements result in further uncertainty in the supplier's business environment. Also, environmental requirements may vary for buyers such as Inditex and H&M, as they have different requirements in addition to a few commonalities. Every buyer has a minimum requirement without which they will not be able to do business with them and that is often not the same as other buyers. Therefore, suppliers need to address buyer-specific standards in each area such as waste management, water quality, chemical usage, harmful chemical, sludge management to continue business with their buyers.

4.2 Variety in Requirements

Bangladeshi garment manufacturers use sandblast technology (dry process) for treating garment colours, which is now under restriction by most buyers. A less hazardous process called potassium permanganate was recommended instead of sandblasting but it is a different process (wet process). There is no certainty that this will continue, in fact, likely to be changed soon. This is because the new regulations and restrictions may come into effect so that process becomes less hazardous to the environment. So, the changes are happening quite frequently. In addition to frequent changes, many requirements are not similar to previous ones. Thus, coping with the variety in the portfolio of requirements becomes challenging for suppliers. To be specific, although manufacturers have LEED certificates (Leadership in energy and environmental design—Platinum, Gold, Silver, etc.) many buyers ask for other certificates such as GOTS which is extra pressure for them.

4.3 Complexity in Implementation

While implementing environmental requirements from buyers, suppliers find them difficult to address and significant complexity involved in their operations. A simple example could be if an environmental requirement demands new methods or additional processes that might be challenging and not realistic within the existing setup. Additional set-up may consume a lot of time, also they might not have the expertise or workforce to fulfil the requirements. Often, they also need to invest separately, which creates many complexities in coordinating organisational functions such as some of the environmental requirements demanding additional spaces, changing to new technologies which will replace existing processes and technologies. Moreover, because of the changing product-based regulations around the world, especially in Europe, and since the Bangladesh market is dominated mainly by European buyers, suppliers face further strict environmental requirements from their buyers, which have complex implementation implications. For example, Zero Discharge of Hazardous Chemicals (ZDHC) is a treaty or programme implemented by worldwide apparel and footwear brands that requires no hazardous chemicals to be discharged by 2020. Suppliers in Bangladeshi manufacturers need to comply with this because many of their buyers signed this treaty; however, they are not aware of its complex implementation. Zero discharge means, recycle 100% of the chemicals used and separating residuals from the process, which is very expensive. Currently, the recycling rate is less than 20% of the water used while they must achieve 100% by 2020. It requires huge investment, reform, and expansion of processes and activities and is thus very complex to implement.

4.4 Newness in Environmental Requirements

In addition to buyer environmental requirements that are difficult to implement, supplier firms also encounter certain new requirements and therefore have no experience and idea of how the requirements might affect their current operations (Shumon et al. 2019). In some cases, it may not necessarily involve huge investments, but suppliers will be less interested to invest in them because the potential impact is uncertain. One example is, as part of the environmental sustainability initiative, many buyers in the Bangladesh apparel market stopped using a dye that is not environmentally friendly. As a result, suppliers need to use a new dye instead. This is new for Bangladeshi manufacturers, and it is uncertain how the existing production process will be impacted. These new requirements not only are limited to product and process but also may include changes in major existing systems. For example, most of the apparel manufacturers in Bangladesh do use biochemical processes in their 'Effluent Treatment Plant' (ETP) to treat wastewater before discharging to the environment, however, big buyers are asking to change it into a *biological process* which was a completely new system for them to implement.

4.5 Timeline

Because of the short life cycle of apparel products, apparel manufacturers face time pressure inevitably. It could also be true for other industries as well for many other reasons such as buyer's opportunism as mentioned by Lee et al. [16]. For example, buyers who have signed the treaty are supposed to comply with changes in environmental regulations in their country within the specific deadline, and all that deadlines are passed to the suppliers to meet and in many circumstances, the timeline given is even before the original deadline. This is because buyers want to keep some days in their hands to be safe in case suppliers cannot meet their deadline. When these requirements are passed to other suppliers upstream in the supply chain, each of them keeps some days in their hand which makes the deadline for the last supplier extremely difficult to meet which is similar to the bullwhip effect and termed as the 'green bullwhip effect' by Lee et al. [16].

4.6 Buyer Specific Framework

Individual buyer has their approach for implementing environmental requirements. Some buyers have their institutional framework and specific set of guidelines for their suppliers to follow. It is challenging for the suppliers to work simultaneously with various institutional approaches. For example, even if a supplier reduces the environmental impact by standard measures, may not be accepted by their buyers; instead, they may need to follow supplier guidelines from each buyer. Often, such institutional frameworks are less applicable for future use, which eventually affects their economies of scope.

4.7 Risk of Penalty

Supplier companies (e.g. ready-made garment suppliers in Bangladesh) face uncertainty when receiving a variety of environmental requirements from their buyers around the world. This uncertainty creates further complexity in suppliers' implementation and operation plans when the introduction of product-based environmental regulations or changes in buyers' environmental strategies are not known to suppliers. Suppliers face stringent timelines and the possibility of penalty from their buyers should they fail to meet buyer's deadlines. There are two types of penalties such as indirect penalty and direct financial penalty. Because of these penalties, suppliers feel pressure to meet the requirements. A direct financial penalty is less likely, however, but lower ratings and points are common through which buyer maintains a list of preferred suppliers. This means, if suppliers cannot meet their buyer's requirements, future order quantity might be less than usual. Therefore, whenever the buyers ask their suppliers to do something regarding sustainability initiatives, they have to address it. Otherwise, the buyer will move to other suppliers who comply with environmental sustainability initiatives of the buyer even if the quality, delivery is within the requirements.

5 Adopting Stringency

Stringency may arise in the context of a dynamic business environment. Therefore, to adopt dynamic changes in environmental requirements, businesses need to prepare themselves accordingly. Researchers suggest that to adopt the unpredictable changes buyers need to develop their capability so that they can absorb the changes. In addition, building good relationships may help them build their capability even in some cases reducing uncertainty. The below section discussed two important aspects of how stringency can be mitigated in the context of the ready-made garment industry.

5.1 Developing Dynamic Environmental Capability

Literature has revealed the importance of suppliers' capability on environmental management issues and focused on to what extent suppliers can improve their performance on environmental issues. Out of the two types of capabilities stated in the literature, most studies mentioned static capabilities which are mainly resource-based. A separate stream of research has argued that firms need to have dynamic capabilities to address stringent environmental requirements. Taking the dynamic capability perspective of Zahra and Jorge [43], it has been argued that in an uncertain business environment where environmental changes are frequent, static capabilities are not able to cope with the changes. Firms need dynamic capability instead to absorb the changes quickly. The dynamic capability framework [43] supports this assumption with the view that in uncertain business environments where environmental requirements are changing frequently, 'static' capability is not adequate to address such changes. Instead, to absorb the changes well, firms' capabilities also need to be 'dynamic' [41]. For example, additional expertise may be required to hire [38], special training might be needed to integrate current environmental requirements [29], processes and systems modification may be necessary. Even, in some cases, a completely new system might be required to set up. Therefore, it is obvious that 'buyer environmental requirements' will drive suppliers' capability development for improved environmental performance. Nevertheless, to what extent suppliers need to improve or adjust their capability will depend on the level of stringency of these requirements, buyer–supplier relationship, their existing practice, expertise, etc. These differences in expertise, exposure to various environmental requirements [16], company size and current environmental standards will all contribute to the relationship between perceived stringency and supplier's environmental capability.

A global initiative by leading retailers around the world aimed at developing environmentally friendly garment factories is pushing suppliers to sign in for the Zero Discharge of Hazardous Chemicals (ZDHC) programme for lowering the discharge of hazardous chemicals. The ZDHC programme signatory brands include Adidas, Benetton, Burberry, C&A, Esprit, G-Star Raw, Gap, H&M, Inditex, Jack Wolfskin, Levi Strauss, L Brands, Li Ning, M&S, New Balance Athletic Shoe, NIKE, PUMA and PVH Corporation. The sudden introduction of ZDHC created a lot of complexities in supplier operations, for example, it requires Bangladesh garment manufacturers to further develop their capabilities in different areas. According to the guidelines [44] itself, a lot of improvements were suggested. This includes but is not limited to, environment specialist requirements, providing special training to employees and so on.

Sustainable Action and Vision for a better Environment (SAVE) is another example where major changes were required to be done by suppliers during 2013–2016. The changes were focused on more efficient use of resources in production, reducing waste (material, water and energy) across all operations of the company, and targeting less energy usage, reducing greenhouse gas emissions by 25%. Initiatives for supplier's capability development include training their employees on how to enhance their environmental performance by conserving resources, carrying out on-site assessment and comparing with international standards, creating and executing roadmap for 25% waste reduction target, and implementing an e-learning system containing high-level information from the sustainability guidelines [30].

It is—therefore can be envisaged that as the stringency increases in buyer requirements, suppliers in response are more likely to develop their capabilities. It can be inferred from previous examples that environmental requirements have a positive influence on suppliers' environmental capabilities. Previous examples have shown that suppliers have developed their capability through initiatives such as the Partnership for Cleaner Textiles (PaCT) project, the SAVE project, the ZDHC programme, training programmes supported by buyers, new technological set-ups and 3R (reduce, reuse and recycle) programmes. Besides, because of specific buyer environmental requirements suppliers needed to invest in setting up effluent treatment plants, recruiting specialists in environmental expertise, less hazardous chemicals, the use of organic cotton and the biological treatment instead of waste instead of chemical treatment [2]. These efforts contributed towards suppliers' capability for improved environmental requirements by their buyers.

So, is developing capability the solution for the problem of meeting uncertain buyer requirements? A recent report (published by BGMEA) reveals that previously average water usage was over 250 L/kg of fabric production which has now have been reduced to 100–120 L/kg. In fact, some factories in Bangladesh RMG industry are using 57 L/kg [2] which is a noble example of environmental capability development. The report also reveals an interview with the Head of Sustainability at one of the largest garment exporters in Bangladesh, in which he said about PaCT, 'We have been able to save water, dyes, chemicals, etc.; reduce our costs of production and even contributed to making water available for the local community. In the year 2016, we have been able to save USD 1.6 million.'

5.2   Building Collaborative Relationships with Buyers

As discussed, the factors of stringency in Sect. 4, it can be expected that the relationship between buyers and suppliers has a role to play. Literature has shown the buyer–supplier relationship as a means of reducing operational, reputational risks that arise from unethical and unsustainable practices in the entire supply chain [1]. Supply chain relationship efforts such as collaboration, supplier integration programmes help achieve superior environmental performance across the supply chain [11, 31]. These initiatives may also include waste reduction through joint efforts, innovation for environment and sustainability including economically attractive environmental solutions, implementing new technologies for embracing environmental changes [20]. More specifically, the collaborative buyer–supplier relationship often helps both buyer and supplier firms achieve superior environmental performance. Some examples from the literature are worth mentioning here. Geffen and Rothenberg [9] mentioned that supplier involvement is vital to develop and implement environmentally sound technologies in the context of automotive paint production. A more recent study by Grekova et al. [10] found that environmental collaboration between buyer and supplier impacts firm performance. Other studies such as Simpson et al. [35] found that 'investment' in buyer–supplier relationship influences plays a significant role to motivate suppliers to meet their sustainability goals.

In the context of stringent environmental requirements, the buyer–supplier relationship has several implications. It can be assumed that a good buyer–supplier relationship reduces suppliers' perceptions of stringency. For example, buyers can provide the latest information about the changes in environmental requirements so that suppliers can prepare in advance and manage the changes. Generally, the RMG sector receives harsh requirements more frequently these days. For example, Zero Discharge of Hazardous Chemicals (ZDHC) was an ambitious target for Bangladeshi garments manufacturers to be achieved by 2020. Usually, buyers employ the penalty mechanism to keep suppliers on track.

In the context of Bangladesh RMG industry, the buyer–supplier relationship is considered a key element for developing environmental capability. Mckinsey and Company [22] also suggested in their report that long-term relationship plays a key role in developing capability not only in the environmental area but also in other areas of the organisations. They also highlighted the high-level investment, and uncertainty associated with it deters Bangladeshi suppliers to address environmental issues. ISO 14001 was the major requirement for Bangladeshi garment manufacturers, many suppliers invested accordingly on developing capabilities such as acquiring environmental standards. But a few years later, now, LEED (even the highest category— LEED platinum) is the prime environmental requirement from buyers. Because of the changes in requirements, long-term relationships are the key for the continuation of contracts and making a return on investment related to environmental capabilities. SAVE, PACT, the Cleaner Production Initiative (CPI) and ZDHC are some examples of buyer's initiatives to build long-term relationships with Bangladeshi suppliers. These examples substantiate the claim that 'relationship' influences the extent of environmental requirements suppliers to receive from buyers. Suppliers who are successful in managing environmental requirements tend to have a trusting relationship and strategic partnership with their buyers. Understandably, as buyers are aware of the supplier's capability, therefore, '*buyers can easily depend*' on their suppliers. In such cases, they collaborate with their suppliers in many aspects such as anticipating and making suppliers aware of any forthcoming changes in environmental requirements, making advance plan and passing them to suppliers, coordinating and managing the changes together instead of directing suppliers in an arm-length relationship. Therefore, a good buyer–supplier relationship may minimise the effect of stringent environmental requirements.

6 A Case Study in Bangladesh RMG Industry

Two garment manufacturers from Bangladesh were identified as Case A and Case B to understand how they perceive stringency and how it impacts different aspects of the environmental performance of the companies. A summary of findings from the case study is presented in Table 1.

In response to the questions regarding the environmental requirements, both companies responded that environmental requirements from buyers change frequently. For example, buyers ask for different types of environmental standards from time to time and that does not last long. In addition, different buyers ask for different environmental standards. To meet that they need to adjust, invest, hire consultants, arrange training programmes and so on. The first point we note is that, in many circumstances, supplier firms can't predict what are the environmental requirements will be in near future. For instance, supplier A said that one of the major buyers asked them to reuse 20% of the wastewater in the current year which was unexpected for them. They were not aware of this kind of forthcoming requirement and also were not informed in advance about this requirement. Further, they are also unsure about next year's requirements, thus they are always in an uncertain situation. Moreover, due to the frequent changes in environmental requirements, more uncertainty arises in the supplier's business environment. Considering the case of supplier B, they explain that requirements are not fixed rather they vary for buyers. They talked about the buyers such as Inditex, H&M and other buyers, most of them have different requirements. Each of the buyers has its minimum standards to continue business with them. Nevertheless, being a preferred supplier requires them to maintain every buyer's specific standards in various areas such as water quality, chemical usage, harmful chemical, sludge management, waste management. On a similar note, Supplier A indicated that they were asked to implement a change in technology for treating garment colour which is a different process and needs some process modifications. Even this is likely to be changed soon because the buyer wants to be even less hazardous to the environment. Therefore, they observe constant changes in environmental requirements.

They also indicated that they face variations in their requirement portfolio which increases their dedicated spending for the specific buyer. This is often challenging for them as they mentioned an example of Global Organic Textile Standard (GOTS) which further added variation in their portfolio. Though they already have LEED platinum (already comply with several environmental standards), still require GOTS only for that specific buyer. To ensure the organic cotton has been used in every stage, they need to go to the end of the supply chain; this is an added pressure for the

Table 1 Summary of findings from the case study

	Stringency	Implication on firm performance (environmental/operational)	Policy/operational changes due to stringency
Case-A	• The requirement from the buyer was to implement ZDHC meaning no hazardous chemical will be discharged to the environment by 2020 • Byer asked them to recycle 100% of the chemicals used and separate residuals which are very expensive and need a lot of investment while currently, they are doing only 10–15% • Buyers require them to meet air quality level which is higher than local guideline (AQI < 100, satisfactory in Bangladesh) • The buyer asked for additional containment of hazardous chemical storage (which is 110% of the original storage) attracted huge investment • Different buyers' acceptance levels for using hazardous chemicals are different (varies from 100 to 50 ppm) • A different list of hazardous chemicals from different buyers creates complexity • Buyers use reduction of order volume as a penalty • One buyer had put a restriction on using sandblast technology	• Reduced energy and electricity usage significantly • Removing sandblast technology from the process improved the health and safety conditions of the production floor • More usage of green chemicals, less water pollution and more recycling efficiency • The new waste management plan treated 15% of their waste last year; this year is getting even better • 100% recycling requires a huge initial investment. They have concerns over the return on investment • Unsure about the acceptance of the wet processing technology by the buyers; therefore, unsure whether the investment will be recovered	• Training on implementing ZDHC • Bought new technology to remove hazardous chemicals • Assigned environmental specialist (ECR or environmentally campaign responsible) to work dedicatedly for a specific buyer • Investment on the latest technology for achieving ZDHC • Modified major processes (non-disclosed) that can now handle a variety of chemicals better than before • A new waste management plant is launched to meet recycling requirements • Implemented wet processing technology (uses potassium permanganate) to replace sandblast (dry) technology

(continued)

Table 1 (continued)

	Stringency	Implication on firm performance (environmental/operational)	Policy/operational changes due to stringency
Case-B	• The buyer asked for using 100% organic cotton • Required to remove hazardous chemicals from the process of (such as lead, phthalate) of sophisticated wash garments, which they could not comply with upfront • A 'dye' was banned while they had a 6-month stock • Working with new chemicals created uncertainty. It took significant time to set up the process again • A tight timeline of only three months was provided for the documentation, sourcing and testing of the new chemicals • The requirement from H&M itself induced complicatedness in the process because of the 150 different styles, colours and codes • A dedicated PaCT program was implemented only for H&M which is of no use for other buyers • Most of the major buyers had their specific frameworks of environmental requirements • The buyer asked to develop a new water and electricity metering system	• Reduced the use of hazardous chemicals in processes and products • Reduced raw materials and water consumption • Improvement has shown a reduction of solid waste, landfilling hazardous liquid and total energy use • New dye was introduced which improved environmental performance, although they had a six-month stock. It is a huge financial loss • Return On Investment (ROI) is a concern as a lot of the capabilities were developed as specific investments and not volume justified	• The source in the supply chain needed to be re-established in response to the 100% organic cotton requirement. Employed a dedicated team to reach overseas farmers • Established SOP and dedicated manpower to address the requirements • Developed new facilities for 'washing' and 'dyeing' • Improved their facilities to maintain indoor air quality index such as temp 280 C, humidity 60%, etc. • Developed a water and electricity metering system to meet H&M requirements

suppliers. Therefore, it can be said that although some environmental requirements are common, many requirements are unique and are not required by other buyers.

Because of the unanticipated changes in the environmental requirements, both suppliers had to push themselves to comply with these changes. For example, stricter containment of hazardous chemicals was required and Zero Discharge of Hazardous Chemicals (ZDHC) programme (which is stricter) was suddenly introduced by some buyers, and Supplier A had to respond to that. As previously mentioned, sandblast technology was also restricted, meaning the current process suddenly became obsolete. As a result, firms were required to expand their present capabilities further, as the manager from Supplier A said they had to make significant changes in their operations to meet the new requirements. They also recruited an expert dedicated to managing their environmental programme, training their employees in homes and abroad, and improving their production process to reduce the level of hazardous chemicals. In addition, they are also using more environmentally friendly techniques such as wet processing (using potassium permanganate) instead of traditional technologies that are more hazardous to the environment. This enables them to manage most of their buyer requirements successfully. Nevertheless, as mentioned before they are uncertain for how long the buyer will accept it. In a similar context, Supplier B was asked to ensure using 100% organic cotton across their supply chain. To ensure 100% organic cotton, more efforts were needed in monitoring and coordination in different stages of the supply chain. Furthermore, the firm installed new monitoring systems for tracking energy (e.g. water, electricity, gas) usage while also establishing new facilities for 'washing' and 'dyeing' more friendly to the environment. They also went through key changes in the firm's capability. One of its major buyers required the firm to comply with an environmental performance improvement project called 'SAVE', which set a higher target to decrease emissions by 15–20% at the end of the year. It essentially focuses on decreasing all sorts of waste (material, water and energy) from the process. Due to 'SAVE', factory-wide lighting systems had to change from traditional to energy-efficient ones by employing modern and costly energy-efficient technologies. This requirement came from a major buyer, so Supplier A had to act quickly with attention to detail. For instance, they recruited a dedicated environmental specialist and developed a comprehensive plan to gradually transform the existing system into the one required by their buyers. The firm was still in the process of executing some requirements.

It is worth noting that both companies are concerned about the 'investment' they had to make for the initiatives. For example, supplier B mentioned that they had to set energy-efficient measures and automated cutting technology for waste reduction in the fabric cutting process which attracted massive investments. However, such investment could not result in more orders or any other form of rewards so they were unsure when they will get the return on investment. Similarly, Supplier A mentioned that a 'biological waste treatment system' was required by one of the buyers and they are spending a great deal of money for that. In addition, to fulfil the requirement of another buyer, they are purchasing more energy-efficient machines/technologies. Although their profit margin has been reduced, however, they are dependent on buyer orders to run the company smoothly. Therefore, few options remain for them but to address the requirements.

Furthermore, both companies have stressed the ongoing relationship with their buyers for the implementation of environmental practices. They mentioned that the relationship is vital for two reasons. First is, they need to know how long the changes they make will sustain, if their relationship is good, buyers often share their change of plans so they can prepare beforehand. Second is, they need help from buyers in many instances, for example, developing capability through collaborative training programmes, processes development, materials and other innovative solutions to environmental problems [29, 33]. They mentioned that their relationships with buyers helped them to improve their capabilities concerning the environmental performance requirements. Specifically, Supplier A admitted that they were able to maintain good relationships with some of their buyers for more than twenty years now. Buyers have not only provided directions and guidance on meeting environmental requirements but also monitored and assessed their performance regularly.

However, the other case has shown opposite views. Specifically, buyers have shown a lack of interest in investing in suppliers' capability to solve environmental issues. Supplier B points out that their buyers think suppliers are responsible for the required capability to meet current environmental requirements. Buyers mostly care about the outcome when assessing their environmental performance rather than a collaborative approach. However, they acknowledged that some buyers regularly monitor their performance and, in some cases, offer support but that's the best they do.

It is evident from case studies that often a buyer's involvement in environmental practices could be based more heavily on the level of relationship they build with their suppliers. For example, a supplier mentioned that their major buyer has helped them to improve their environmental performance through a supplier development programme. They arrange frequent visits to run relevant training programmes at supplier premises and at times they also provide specialist help when required.

7   Conclusion

The concept of 'stringent' customer environmental requirements is relatively new in the sustainable supply chain domain. Environmental requirements from buyer firms are inevitable because of the sustainability pressure. However, the perspective of supplier firms concerning these requirements is not well established yet in the literature. Bangladesh RMG industry is a good example where supplier firms also face these stringent requirements across the buyers. This study is an endeavour to draw this concern which needs to be addressed in academic arenas and find out the ways how supplier firms can handle and respond to these requirements properly. Two possible solutions were recommended in this study to mitigate stringency in the environmental sustainability context. They are the firm's environmental capability and the relationship between buyers and suppliers. Also, this study emphasises the need for further research on these two broad-level solutions, for example, the challenges for environmental capability development and long-term relationships for sustainability.

References

1. Anne T, Helen W (2015) Theories in sustainable supply chain management: a structured literature review. Int J Phys Distrib Logist Manag 45(1/2):16–42
2. BGMEA (2017), The apparel story, URL: http://www.bgmea.com.bd/media/newsletters/May-August-2017/May-August-2017.pdf.
3. BGMEA (2021) Bangladesh home to world's highest number of green garment factories, URL: https://www.bgmea.com.bd/index.php/page/Sustainability_Environment
4. Bowen FE, Cousins PD, Lamming RC, Farukt AC (2001) The role of supply management capabilities in green supply. Prod Oper Manag 10:174–189
5. Carlton J (2006) 'EU's environmental hurdles for electronics; rules to require mitigation of toxic materials that are common in many products. June 29, B5', Wall Str J, June 29
6. Darnall N, Henriques I, Sadorsky P (2010) Adopting proactive environmental strategy: The influence of stakeholders and firm size. J Manage Stud 47(6):1072–1094
7. Delmas MA, Montes-Sancho MJ (2010) Voluntary agreements to improve environmental quality: Symbolic and substantive cooperation. Strateg Manag J 31(6):575–601
8. Economist (2015) The growing environmental costs of a Brazilian disaster. https://www.economist.com/the-americas/2015/11/27/the-growing-environmental-costs-of-a-brazilian-disaster
9. Geffen CA, Rothenberg S (2000) Suppliers and environmental innovation: The automotive paint process. Int J Oper Prod Manag 20(2):166–186
10. Grekova K, Calantone RJ, Bremmers HJ, Trienekens JH, Omta SWF (2016) How environmental collaboration with suppliers and customers influences firm performance: Evidence from Dutch food and beverage processors. J Clean Prod 112:1861–1871
11. Gold S, Seuring S, Beske P (2010) Sustainable supply chain management and inter-organizational resources: a literature review. Corp Soc Responsib Environ Manag 17(4):230–245
12. Green K, Morton B, New S (1998) Green purchasing and supply policies: do they improve companies' environmental performance? Supply Chain Manag: Int J 3(2):89–95
13. Halldórsson Á, Kovács G, Mollenkopf D, Stolze H, Tate WL, Ueltschy M (2010) Green, lean, and global supply chains. Int J Phys Distrib Logist Manag 40(1/2):14–41
14. Ken G, Barbara M, Steve N (1998) Green purchasing and supply policies: do they improve companies' environmental performance? Supply Chain Manag: Int J 3(2):89–95
15. Lee HL, Padmanabhan V, Whang S (1997) The bullwhip effect in supply chains. Sloan Manag Rev 38(3):93–102
16. Lee S-Y, Klassen RD, Furlan A, Vinelli A (2014) The green bullwhip effect: Transferring environmental requirements along a supply chain. Int J Prod Econ 156:39–51
17. Lee S-Y, Klassen RD (2008) Drivers and enablers that foster environmental management capabilities in small- and medium-sized suppliers in supply chains. Prod Oper Manag 17(6):573–586
18. Liu Y, Zhu Q, Seuring S (2017) Linking capabilities to green operations strategies: The moderating role of corporate environmental proactivity. Int J Prod Econ 187:182–195
19. Lu S (2018) Market size of the global textile and apparel industry: 2016 to 2021/2022. Accessed on 06/04/2018, Link: https://shenglufashion.com/tag/textile-industry/
20. Matos S, Hall J (2007) Integrating sustainable development in the supply chain: The case of life cycle assessment in oil and gas and agricultural biotechnology. J Oper Manag 25(6):1083–1102

21. Melnyk SA, Sroufe RP, Calantone R (2003) Assessing the impact of environmental management systems on corporate and environmental performance. J Oper Manag 21(3):329–351
22. Mckinsey and Company (2011) Bangladesh: The next hot spot in apparel sourcing? Viewed 28 June 2018 <http://www.bgmea.com.bd/beta/uploads/pages/2011_McKinsey_Bangladesh_Case_Study.pdf>
23. Meixell MJ, Luoma P (2015) Stakeholder pressure in sustainable supply chain management: A systematic review. Int J Phys Distrib Logist Manag 45(1/2):69–89
24. Montiel I, Husted BW (2009) The adoption of voluntary environmental management programs in Mexico: First movers as institutional entrepreneurs. J Bus Ethics 88(2):349–363
25. Murillo-Luna JL, Garcés-Ayerbe C, Rivera-Torres P (2008) Why do patterns of environmental response differ? A stakeholders' pressure approach. Strateg Manag J 29(11):1225–1240
26. Obe M (2018) Bangladesh fights for future of its garment industry, viewed on 06/04/2019 link: https://asia.nikkei.com/Business/Business-trends/Bangladesh-fights-for-future-of-its-garment-industry
27. Prothom-Alo (2016) Seven companies from Banlgaldesh in the list of top 10. Prothom-alo, Accessed on 03/12/2016
28. Rueda-Manzanares A, Aragón-Correa JA, Sharma S (2008) The influence of stakeholders on the environmental strategy of service firms: the moderating effects of complexity, uncertainty and munificence. Br J Manag 19(2):185–203
29. Sarkis J, Gonzalez-Torre P, Adenso-Diaz B (2010) Stakeholder pressure and the adoption of environmental practices: The mediating effect of training. J Oper Manag 28(2):163–176
30. SAVE (2018) SAVE Projetc report' URL: http://about.puma.com/damfiles/default/sustainability/SAVE/SAVE_Final_Report_20161013.pdf-24bdf8142d1f74ecf13e3a70302a22c2.pdf. Accessed on 20/062018
31. Seuring S, Müller M (2008) From a literature review to a conceptual framework for sustainable supply chain management. J Clean Prod 16(15):1699–1710
32. Simpson D, Power D (2005) Use the supply relationship to develop lean and green suppliers. Supply Chain Manag: Int J 10(1):60–68
33. Simpson D, Power D, Samson D (2007) Greening the automotive supply chain: a relationship perspective. Int J Oper Prod Manag 27(1):28–48
34. Simpson D (2012) Knowledge resources as a mediator of the relationship between recycling pressures and environmental performance. J Clean Prod 22(1):32–41
35. Simpson D, Power D, Klassen R (2012) When one size does not fit all: A problem of fit rather than failure for voluntary management standards. J Bus Ethics 110(1):85–95
36. Sony (2015) Road to Zero: Sony's Global Environmental Plan, viewed 24/11/2015. http://www.sony.net/SonyInfo/csr_report/environment/management/roadto/index.html
37. Starbucks (2015) Ethical sourcing. viewed on 24/11/2015; Link: http://www.starbucks.ca/responsibility/sourcing
38. Tseng SM, Lee PS (2014) The effect of knowledge management capability and dynamic capability on organizational performance. J Enterp Inf Manag 27(2):158–179
39. Viyellatex (2018) Corporate Sustainability report 2017, http://www.viyellatexgroup.com/wp-content/uploads/2018/02/Viyellatex-Sustainability-Report-2017.pdf
40. Walker H, Di Sisto L, McBain D (2008) Drivers and barriers to environmental supply chain management practices: Lessons from the public and private sectors. J Purch Supply Manag 14(1):69–85
41. Winter SG (2003) Understanding dynamic capabilities. Strateg Manag J 24(10):991–995
42. Wong CW, Lai KH, Shang KC, Lu CS, Leung TKP (2012) Green operations and the moderating role of environmental management capability of suppliers on manufacturing firm performance. Int J Prod Econ 140(1):283–294
43. Zahra SA, George G (2002) Absorptive capacity: A review, reconceptualization, and extension. Acad Manag Rev 27(2):185–203

44. ZDHC (2018) 'Chemical Management Systems Guidance Manual' https://www.roadmapto zero.com/fileadmin/layout/media/downloads/en/CMS_EN.pdf. Accessed on 20/06/2018
45. Zhu Q, Sarkis J (2007) The moderating effects of institutional pressures on emergent green supply chain practices and performance. Int J Prod Res 45(18–19):4333–4355

Immobilization as Sustainable Solutions to Textiles Chemical Processing

Amit Madhu

Abstract Sustainability has become an integral part of textile manufacturing industries in recent years. Enzymes are one of the most sustainable alternatives and advancement in biotechnology has developed more tailored enzymes for various textile process applications. Enzymes have already found commercial success in textile processing, and several life cycle assessment (LCA) studies have evident enzymes as a promising approach to reducing pollution, conserving resources, and lowering costs. The native enzymes, however, do not satisfy the criteria for large-scale use. One of the most promising techniques for highly efficient and economically competent biotechnological processes is enzyme immobilization. Immobilization of enzymes is a valuable method for effective recovery and reuse of expensive enzymes, as well as better enzyme function through improved stability in both storage and operating settings. Reduce, reuse, and recycle are the core tenants of sustainability; thus, immobilized enzymes can be a real sustainable approach for the bio-processing of textiles. In another potential application, the immobilization of enzymes on textiles can add additional functionalities to textile. A few naturally occurring enzymes have recently been discovered to have the potential to be implemented as biological protective finishes after immobilization on textiles. Furthermore, textile materials can serve as sustainable support materials for immobilization, and biocatalysts immobilized textile opens up exciting possibilities for developing a reliable fiber-based catalytic system for a variety of industrial-scale applications.

Keywords Biocatalysts · Bio-processing · Enzyme · Immobilization · Life cycle assessment · Sustainability · Textile · Wastewater

A. Madhu (✉)
Department of Textile Chemistry, The Technological Institute of Textile and Sciences, Bhiwani (Haryana), India
e-mail: amitmadhu@titsbhiwani.ac.in

© The Author(s), under exclusive license to Springer Nature Singapore Pte Ltd. 2022   21
S. S. Muthu (ed.), *Sustainable Approaches in Textiles and Fashion*,
Sustainable Textiles: Production, Processing, Manufacturing & Chemistry,
https://doi.org/10.1007/978-981-19-0538-4_2

1 Introduction

Enzymes are extremely efficient biocatalysts researched for industrial-scale catalysis because of several distinct benefits ranging from their ability to operate in gentler reaction conditions to their remarkable product selectivity and reduced environmental and physiological toxicity. They have been efficiently used as biocatalysts for many chemical processes in the pharmaceuticals, diagnostics, biosensors, food and beverage, detergent, and bio-fuel sectors due to their decreased energy needs, waste reduction, and simpler manufacturing methods [15, 85]. In the textile industry enzymes are found to be applied to all manufacturing steps while their extensive application is mainly in wet processing, detergent formulation, development of biodegradable fibers, biosynthesis of dyes/colorants, and the treatment of textile effluent [8, 18, 32, 86].

Enzymes are widely employed in textile chemical processing because of their ability to substitute nasty chemicals and save water and energy. However, exorbitant prices and a lack of long-term stability under storage and process conditions stymied their industrial commercialization. Because enzymes are not consumed in processes as biocatalysts, reusability may be the most cost-effective option. Recent advances in protein engineering have revolutionized the transformation of commercially accessible enzymes into more effective industrial catalysts. Immobilization of enzymes is one such method for enhancing enzyme characteristics [15, 50, 86]. Today, the extensive research work is associated with the immobilization of various enzymes for industrial applications, to expand their horizon for textile applications to achieve sustainability.

This chapter summarizes the current status of enzyme technology as a sustainable approach and emphasizes breakthroughs associated with the application of immobilized enzymes in textile chemical processing. This also includes a brief overview of the fundamentals of enzymes immobilization including immobilization techniques, support materials, and performance of immobilized enzymes. Further, this chapter contains an in-depth discussion on the latest enzyme immobilization techniques with various applications of immobilized enzymes in textile wet processing. The immobilization of enzymes for textile functionalization and the concept of "biocatalysts immobilized textiles" are also reviewed along with their potential application in the discoloration of textile wastewater.

2 Need of Sustainability in Textile Chemical Processing

Textiles manufacturing is one of the oldest industries and it contributes significantly to the growth and economy of many countries. Mainly in developing countries, the textile and garment industry is a key source of manufacturing and trade, being the availability of cheap personals and weak environmental legislation. Textile chemical processing is an important step in textile manufacturing as it adds aesthetics, comfort, and functional properties to textiles. Conventional textile processing includes numerous chemical and non-chemical treatment processes from pretreatments to their final finishing. These processes use a lot of chemicals and water, which are then discharged as effluents into the environment [18, 73, 86].

Textile chemical processing is the second most polluting industry in the world after the chemical manufacturing industry. Despite being a major source of revenue, the textile sector has become a major polluter across the world, posing substantial environmental risks. During chemical processing, textile materials are exposed to a variety of hazardous chemicals; these chemicals are expensive and pose a substantial threat to the environment, ecosystem, and aquifers, as well as worker health. Many auxiliaries used in chemical processing, such as wetting agents, dispersion agents, and surfactants, are non-biodegradable and non-recyclable [47]. The majority of dyes used to color textiles are synthetic, non-biodegradable, and made from aromatic compounds, and these dyeing colorants do not get fixed and end up as effluent. Formaldehydes, fluorinated polymers, triclosan, brominated, or antimony-oxide-based flame retardants are used in textile finishing and are not only poisonous but also carcinogenic in some cases, and are on the list of restricted substances in many countries [32, 61, 82].

The textile industry is also one of the major consumers of water; a large quantity of freshwater is used as a medium in various wet processes starting from pretreatments (desizing, scouring, and bleaching) to dyeing, printing, and value-added aesthetic and functional finishing. Along with that a considerable amount of water is also required for soaping and rinsing of textiles materials, as well as for the creation of steam to heat the process bath. The textile industry consumes over 1.5 trillion gallons of water each year, with the amount of water required for processing varies depending on the type of material, chemicals, and auxiliaries used, and processing machinery; on average, 50–100 L of water is required to process one kilogram of textile material (economicsofwater.weebly.com [63, 82], sustainablecampus.fsu.edu).

All of the water used in textile processing is expelled at the end, with an average of 90–95% of the water used ending up as effluent. This discharged water is highly toxic and is a cocktail of unfixed dyes, residual chemicals, auxiliaries, etc. As a result, textile industry effluent is a complex mixture with high biological oxygen demand (BOD), chemical oxygen demand (COD), total dissolved solids (TDS), pH, low dissolved oxygen, and intense color [61, 64]. Thus this effluent generated would require special treatment to bring down these parameters within the permissible limits before discharge. As most of the developing countries are associated with the chemical processing of textiles and they lack in resources to adopt modern expensive effluent treatment techniques, they discharge effluent into the environment without sufficient removal of toxic residuals. The ecological and toxicological hazard caused by the discharge of textile industry effluents into streams and other inland water bodies is a serious concern across the world. Approximately, 20% of industrial water pollution comes from textiles treatment and dyeing (economicsofwater.weebly.com). Increased global textile demand has resulted in a freshwater constraint for processing, as well as pollution of water bodies, emphasizing the issue of textile chemical processing sustainability [64, 82].

Furthermore, textile processing consumes a lot of energy in heating, drying, and running the machines, which adds considerably to greenhouse gas emissions and carbon footprints [22, 32]. Chemical processing consumes about 21–69 MJ/kg of energy from pretreatment to finishing, whereas total textile manufacturing consumes approximately 125.05 MJ/kg of energy [5]. Although electric energy is the most commonly utilized form of energy in textile production, thermal energy uses the bulk of energy in chemical processing after electric energy. Because many textile manufacturing processes are carried out at high temperatures, therefore energy is required for heating the bath and for drying textiles in intermittent stages. Dyeing consumes most of the energy in the textile industry in India, accounting for 25.6% of energy consumption. This thermal energy is frequently provided by steam produced by the combustion of fossil fuels, which produces greenhouse gases and leaves significant carbon footprints [61, 63, 64] (sustainablecampus.fsu.edu).

In addition, indirect carbon footprints are linked with textiles due to the inherent energy of chemicals and raw materials. Textile fibers have diverse carbon footprints as raw materials, yet they have similar carbon footprints after processing [62, 64]. In a completely continuous textile finishing process for cotton textiles, over 50% of emissions originate from drying, 40% from washing and steaming, and 10% from chopping. Thus, massive consumption of water, energy, carbon footprints, hazardous chemicals, and the disposal of noxious effluent are the primary problems in the long-term sustainability of textile wet processing operations [5, 37]. Figure 1 gives the

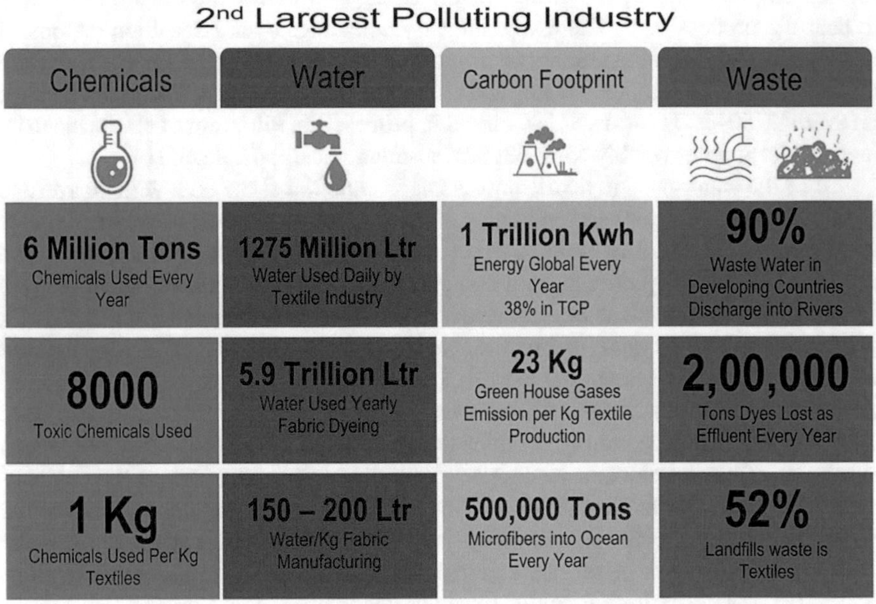

2nd Largest Polluting Industry

Chemicals	Water	Carbon Footprint	Waste
6 Million Tons Chemicals Used Every Year	**1275 Million Ltr** Water Used Daily by Textile Industry	**1 Trillion Kwh** Energy Global Every Year 38% in TCP	**90%** Waste Water in Developing Countries Discharge into Rivers
8000 Toxic Chemicals Used	**5.9 Trillion Ltr** Water Used Yearly Fabric Dyeing	**23 Kg** Green House Gases Emission per Kg Textile Production	**2,00,000** Tons Dyes Lost as Effluent Every Year
1 Kg Chemicals Used Per Kg Textiles	**150 – 200 Ltr** Water/Kg Fabric Manufacturing	**500,000 Tons** Microfibers into Ocean Every Year	**52%** Landfills waste is Textiles

Fig. 1 Environmental impact of textiles chemical processes

data to emphasize the importance of more sustainable and transparent practices in textile chemical processing.

3 Sustainability in Textile Chemical Processing

Sustainability is a multidimensional notion that includes environmental, economic, and social implications. It is described in the manufacturing sector as the production of non-polluting goods that conserve energy and natural resources, are financially viable, and are safe for employees, communities, and customers. The environmental component is based on demands for reduced natural resource usage and waste, as well as environmental protection. The economic dimension of sustainability considers aspects including the use of sustainable raw materials, the total cost of ownership, inventory reduction, reuse, and consumption reduction [10, 22, 61, 62, 66].

To achieve sustainability, multiple approaches are currently being adapted and implemented in the textile processing industries. Non-toxic or biodegradable chemicals, well-organized processing with optimized parameters, eco-friendly processing with enzymes, technological advancements in processing machinery, futuristic waterless technologies such as plasma technologies, supercritical carbon dioxide, and digital printing, as well as the use of reusing and recycling techniques can all help textile processing be more sustainable [5]. Enzymes are one of the most environmentally friendly options, and biotechnology advancements have resulted in the creation of more tailored enzymes for a variety of textile processing applications [29, 50, 86].

4 Perspectives on Enzymes in Textile Wet Processing

Enzymes are biological molecules; chemically, they are proteins made up of polypeptide chains, each of which forms a distinct 3-D complex structure. This 3-D structure offers an "active site" for enzymes, similar to the "key-lock mechanism," and both the enzyme and the substrate have complimentary geometric forms that fit precisely into one another. This mechanism demonstrated how biological responses are carried out and the reactive substrate serves as a key that fits perfectly into the enzymatic lock. The most fundamental characteristic of enzymes that makes them excellent for chemical reactions is the specificity of their nature as it relates to the reaction they catalyze. The rate of reaction is catalyzed by most enzymes by factors of 103–1016 when compared to the rate of unanalyzed reaction [85, 86]. The majority of enzymes work on smaller substrates in comparison to their size. Changes in the three-dimensional structure of enzymes can readily denature or render them inactive. The rate of enzymatic processes is influenced by temperature, pH, enzyme concentration, substrate concentration, and any activators, inhibitors, or retarders [8].

Commercial enzymes are mostly derived from animal tissue, plants, and microbes. Fungi provide almost half of the enzymes used in industrial processes, bacteria provide over a third, and animal and plant sources provide the remainder. Due to high output levels associated with a standard expression, ease of growth, low-cost culture medium, and short fermentation cycles, microbial strains generate enzymes that are cost-effective on a large scale. Furthermore, different microbes produce slightly different enzymes that catalyze the same process, providing a lot of versatility [18, 85].

The International Union of Biochemistry (IUB) divides enzymes into six groups depending on the chemical reaction they catalyze, which is the distinguishing feature that distinguishes one enzyme from another. The six types of enzymes are oxidore-ductases (which catalyze oxidation/reduction), transferases (which catalyze transfer a functional group), hydrolases (which catalyze bond hydrolysis), lyases (which cleave bonds), isomerase (which catalyzes the isomerization in a single molecule), and ligases (which join two molecules via a covalent bond). In nomenclature, enzymes are given names based on the substrates on which they react, and all enzyme names end in "ase." Cellulase, for example, acts on cellulose [18, 58, 86].

More than 3,000 distinct enzymes have been identified and categorized too far; however, only a small number are commercially available, and even fewer are utilized in substantial amounts. After detergents and the food industry, textiles utilize the bulk of commercially manufactured enzymes (12%). AB Enzymes (Germany), DUPONT (USA), Novozymes (Denmark), Genencor (Denmark), DSM (Netherlands), BASF (Germany), Nagase (Japan) Kerry Group (Ireland), Aum Enzymes (India) Advanced Enzymes (India) and Chr. Hansen (Denmark) are among the major businesses involved in the manufacture and commercialization of enzymes [15, 18, 86].

The majority of enzymes employed in textile chemical processing are hydro-lases, which include amylases, cellulases, pectinases, catalases, and proteases, and are utilized in a variety of textile manufacturing processes to catalyze the hydrol-ysis of chemical bonds. Enzymatic desizing of cotton fabrics with α-amylases has shown to be successful over time. Bio-scouring for hydrophobic impurities removal; bio-polishing to eliminate fiber fluff from the cotton fabric surface; bio-fading of denim for a fashionable aged effect; bleaching cleanup to remove leftover H_2O_2 before dyeing. Also, efforts have been made to replace traditional anti-shrinking/anti-pilling of wool and degumming of silk with protease enzymes as well as using pectinases, xylanases, or hemicellulases, for retting of bast fibers (jute, flax, ramie) [8, 18, 50, 58].

Furthermore, detergent formulations containing a combination of hydrolase enzymes are used to remove a variety of stains during garment laundry. Amylases and cellulases are well-known enzymes that have been employed in detergents for decades. Mannanases are now used in detergents and cleansers to remove the gummy debris found in many foods and household items. Novozyme sells a mannanase prepa-ration called "MANNAWAY," which is used in laundry detergents. Mannanase can help enhance the whiteness of cellulosic fabrics as well as eliminate stains [30, 91].

Laccases and peroxidases from the oxidoreductase family, on the other hand, have discovered intriguing textile uses such as bleaching textiles, decolorizing of textile effluents, textile surface modification, and in situ dye synthesis. Enzymes can be used in nearly every stage of natural and synthetic textile chemical processing, reducing the usage of polluting and noxious chemicals and resulting in a more sustainable and environmentally friendly textile industry. Table 1 [15, 18, 46, 58, 86] includes several enzymes, along with their mechanisms, that can be used alone or in combination in textiles for various applications. Several enzymes are currently commercially

Table 1 Summary of application of enzymes in chemical processing of textiles

Enzyme class	Name of enzyme	Action mechanism	Textile processing step	Applicable material
Hydrolases	Amylase	Break starch to soluble sugars	Bio-desizing	Cotton and cotton blended textiles
		Remove starch-based stains	Detergent formulation	Any textile laundering
	Pectinase	Decomposition of pectin	Bio-scouring	Cotton and cotton-blended textiles
	Lipase	Decomposition of fats and oils to glycerol and fatty acids	Bio-scouring Surface modification of synthetic fibers	Cotton and cotton-blended textiles Polyester, Polyamides
		Removal of lipid stains	Detergent formulation	Any textile laundering
	Cutinase	Hydrolysis of insoluble cutin	Bio-scouring Surface modification of synthetic fibers	Cotton and cotton-blended textiles Polyester, polyamides
	Xylanase	Degradation of xylan	Bio-scouring Bio-bleaching	Bast fibers: flax, jute, ramie
	Arylesterase	In situ generate peracetic acid Hydrolysis of ester linkage	Bio-bleaching Surface modification of synthetic fibers	Cotton-based textiles Polyester
	Protease	Hydrolysis of peptide bonds of proteins into soluble polypeptides and amino acids	Silk degumming Wool modification	Silk-based textiles Wool and wool-blended textiles
	Cellulase	Breakdown of cellulose into soluble products	Bio-polishing Bio-scouring Bio-stoning (denim fading) Wool carbonization	Cotton and cotton-blended textiles Denim textiles Wool-based textiles

(continued)

Table 1 (continued)

Enzyme class	Name of enzyme	Action mechanism	Textile processing step	Applicable material
Oxidoreductases	Laccase	Degrading of recalcitrant organic compounds through catalyzing transfer of electrons from one molecule Oxidation of flavonoids to prepare dye precursor	Textile wastewater decolorization Bio-bleaching Bio-dyeing (in situ dye synthesis on polymers)	Effluent generated from the textile industry Cotton-based textiles Cotton and protein textiles
	Peroxidase (Catalase)	Breakdown of peroxides (H_2O_2) to water and oxygen	Bleach bath clean-up Textile wastewater Treatment Sulfur dye reduction	H_2O_2 removal before cotton reactive dyeing Effluent generated from textile industry Cotton dyeing
	Glucose oxidase	Generate H_2O_2 on oxidation of glucose	Bio-bleaching	Cotton-based textiles

accessible, and researchers are still working hard to develop new enzymatic products and methods that can be scaled up from the lab to the industry to give the same benefits.

5 Enzymes are Sustainable Solution in Textile Chemical Processing

Since the late 1980s, enzymes have been extensively used in the textile industry and many enzymes have been developed to replace chemicals in textile chemical processing. Enzymes not only assist the environment but also help save money by conserving energy, water, and chemicals while improving quality. They catalyze the reaction by decreasing the activation energy of the process and remain intact at the end of the reaction after the conversion of reactant to product [85, 86]. Enzymatic reactions are several orders of magnitude faster than chemical processes. Because enzyme activity is dependent on operating conditions, processes are simple to manage. These advantages are becoming increasingly relevant as people become more aware of the need for environmental sustainability. A voluminous literature is available for the environmental benefits of enzymatic processing over conventional chemical processing. The many advantages of employing enzymes in textile chemical processing are depicted in Fig. 2.

It may look obvious to believe that enzymes contribute to sustainable development based on a qualitative assessment. However, a quantitative examination of the environmental impact assessment is required to ensure long-term viability. Sustainability in the manufacturing industries means that all the associated materials, processes, inputs, and outputs are healthy and safe for humans and the environment at all stages of the product life cycle and that all energy, materials, and process inputs come from renewable or recycled sources. As a result, to estimate the actual environmental benefits of enzymatic processes, it is also required to investigate the environmental implications of enzyme production [36, 63, 68].

Life cycle assessment or LCA is a methodology for assessing total environmental impacts associated with all the stages of the life cycle of a commercial product, process, or service from its cradle to grave. Each product has a life cycle and a supply chain, and products that move through all of these stages have an environmental impact, such as energy use, water consumption, greenhouse gas emissions, and waste creation. LCA is used to assess the environmental impact of a product's constituent materials, from their extraction and processing to its production, distribution, and use, as well as their recycling or final disposal. LCA gives a more complete picture of the environmental properties of the processes considered than environmental impact assessment (EIA) and carbon footprint analysis. Several researchers conducted LCA studies for a variety of enzymatic procedures to determine the true environmental benefits of enzymatic treatment. The goal of the LCAs is to evaluate and contrast the environmental effects of enzyme synthesis and distribution with the environmental burdens saved in textile chemical processes [10, 22, 64].

A study carried out in China confirms that the application of enzymes in the textile industry reduces water consumption by about 9–14% and energy consumption by 17–18%, during the bio-bleaching process. Additionally, bio-scouring has the potential to reduce water use by 30–50%, while bio-stone washing reduces the expenses associated with water usage and air emissions by 50–60% [18].

LCA research of pectate lyases for scouring and catalases for bleach clean-up found that the impact of enzyme synthesis is minor when compared to the amount

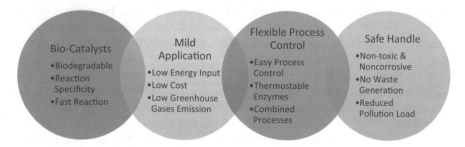

Fig. 2 Advantages of employing enzymes in textile chemical processing

of water, energy, and chemicals saved. Enzymes are thought to have a key role in lowering water use and pollution. In comparison to conventional alkaline scouring, the case study with Scourzyme®301 L (Novozymes) indicates that a little dose (10 kg) of enzyme can save 2.5 tonnes of steam, 150 kWh energy, and 20 m^3 of water per tonne of yarn scouring. Bio-scouring is gentler than traditional harsh alkaline scouring, resulting in a 1.5% yarn weight loss compared to more than 4% in conventional scouring. A small amount of low-toxicity enzyme replaces a larger amount of harsh chemicals, and environmental impact assessment results show that heat savings in bio-scouring, followed by electricity and yarn savings, are the key factors behind the reduced contribution to global warming [36, 68].

An LCA analysis of enzymatic bleach clean-up of knitted fabrics with Terminox Ultra® 50 L (Novozymes) as an alternative to hot water rinsing revealed that the consumption of resources and environmental consequences caused by enzyme production are often insignificant when compared to the savings. The enzymatic bleach clean-up resulted in a considerable reduction of 1820 kg steam, 33 kWh energy, and 20 m^3 water per tonne of knitted fabric processing. Heat conservation is by far the most essential factor in reducing global warming contributions, and the use of enzymes also minimizes environmental toxicity [36].

In a similar life cycle analysis, the environmental implications of traditional and Gentle Power BleachTM (GPB) cotton fabric processing are compared. GPB is an enzyme-based bleaching method (arylesterase) that allows bleaching to take place at lower temperatures and with less water. GPB outperformed conventional in most impact categories, including climate change, human health, ecosystem quality, and water use, by at least 20% [26]. In a study, it was found that when activator agent-assisted hydrogen peroxide bleaching was paired with enzymatic scouring, there were no time or energy savings, but there was a reduction in water and steam consumption, as well as cost [18, 68]. Despite the fact that enzymes were more expensive than conventional materials, the enzymatic methods saved a lot of water and energy and had a lower total cost than alkaline scouring.

Surfactants used in laundry detergents are hazardous, and heating laundry water for activation requires a significant amount of energy. Enzymes are utilized as additives in detergents because they may dissolve stains at low washing temperatures and are less harmful than surfactants. Nielsen and Skagerlind conducted an LCA analysis on a model detergent in which four enzymes (protease, lipase, amylase, and cellulase) were used to replace the three surfactants (ethoxylated alcohol, linear alkylbenzene sulfonate, and sodium soap). The findings reveal that enzyme manufacturing has a minimal environmental impact when compared to surfactant production and that using enzymes saves energy in the use phase while also reducing contribution to aquatic toxicity in the disposal phase [67].

Furthermore, textile manufacturing produces a significant amount of CO_2, not only from the energy used in the business but also from the energy used in the creation and delivery of raw materials, chemicals, and water [63]. Enzymes of various varieties can help textile manufacturers reduce their carbon footprint by preserving raw materials, substituting chemicals, and conserving water and energy during the manufacturing process. Enzymes appear to be one of the most promising approaches for decreasing pollution, preserving resources, and lowering prices as a result. In the future, combining various enzymatic activities in the same bath will help save water and streamline processing, resulting in cost savings [18, 66].

While profit is still the most significant motivator of process development, life cycle assessments (LCAs) and techno-economic analyses (TEAs) combined are crucial tools for properly assessing both the benefits and drawbacks of scaling up biocatalytic processes. LCAs and TEAs are becoming increasingly important methodologies as a growing number of enzymes are being examined for commercial-scale biocatalytic processes. TEAs look at the economic feasibility of a process in terms of technology readiness as well as process economics such as utilities, feed-stocks, labor, and capital inputs [15, 64].

6 Need for Enzyme Immobilization

Enzymes as biocatalysts have proven to be eco-friendly and reliable alternatives to toxic chemicals and have immense numbers of applications in textile chemical processing. However, despite their biocompatibility and beneficial characteristics, the field application is impeded by the limitations associated with their commercial applicability. Some of the challenges of enzyme applications are depicted in Fig. 3.

Thermodynamic and kinetic stability is one of the major constraints in industrial application Their application frequently requires stabilization to several process parameters pH, temperature, reactant concentrations, auxiliaries type and concentration, solvents/co-solvents, and application methodologies like shear and surface forces. The enzymes must be able to withstand extreme conditions such as high temperatures, pH, salts, alkalis, surfactants, and other chemicals commonly used in textile chemical processing. However, modifying the process parameters (e.g., pH, temperature, and auxiliaries) to fit the enzyme's sensitivities can boost process efficiency to some extent. The enzymes are significantly expensive and difficulty

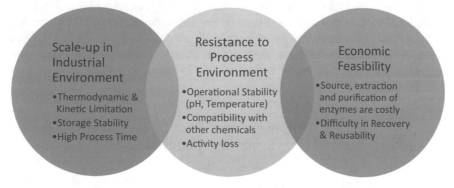

Fig. 3 Challenges in the application of enzymes for textile chemical processing

in recovery and reusability of the enzyme leads to a significant increase in capital investments that restricts its widespread industrial use [46, 50, 69, 92].

The assistance of physiochemical techniques like plasma or ultrasound along with biotechnological enzyme applications enhances the efficiency of enzymatic processes. Ultrasound sonication apart from affecting enzyme activity, more importantly, initiates dominant shock waves that cause actual mixing of the enzyme in heterogeneous substrates systems, such as application of cellulase enzymes on cellulose, pectin hydrolysis by pectinase enzyme, or starch hydrolysis by amylase enzyme. Similarly, the plasma assists in surface modifications to enhance the enzyme action. In all such cases, the rate of enzymatic reactions becomes fast resulting in the reduction of treatment time [81, 85].

Recombinant DNA technology and genetic engineering are being used to generate new enzymes to function under more ideal process conditions (extreme temperatures and pH), as well as biocatalysts with the catalytic activity that are several orders of magnitude larger than those found naturally [15]. While advances in protein engineering have aided in the expansion of industrial enzyme catalysis, but they haven't addressed all of the issues, such as the costs associated with their in vitro production, or their further adoption in commercial-scale processes, poor mechanical stability, and limited reusability. This leads to the development of tailored enzymes that are highly specialized for a range of industrial applications; nonetheless, in all of these circumstances, industrial application of enzymes is still not a cost-effective alternative [44, 50, 69, 92].

Recycling and subsequent reuse of these biocatalysts are extremely desirable from a sustainable and cost-effective standpoint. Several strategies for accomplishing this goal have been proposed during the last few decades. Material scientists and biotechnologists are concerned with immobilizing enzymes or binding them to a support material without losing their essential capabilities. Enzyme immobilization can usually overcome these drawbacks of operation stability, recovery, and recyclability, making them industrially and commercially viable to improve enzyme stability, adaptability to processes, and improved reaction control, as well as simplify biocatalyst recycling and downstream processing [11, 44, 50].

The ability to immobilize and reuse an enzyme catalyst has significant benefits in industrial biotechnology since it decreases industrial investments while improving profit by making the entire process economically possible. While evaluating whether or not to use immobilized biocatalysts on an industrial scale necessitates a thorough examination of both technical and economic issues, as well as the process in consideration.

7 Fundamentals of Enzyme Immobilization

Immobilization of enzymes is referred to as the physical confinement or attachment of enzymes on or to a support material, with retention of their catalytic activities. Immobilization decreases enzyme mobility or confines enzymes within a support

Table 2 Benefits and drawbacks connected with enzyme immobilization

Benefits	Drawbacks
Facile separation of biocatalysts after reaction completion	Enzyme loss during immobilization
Improved stability to the operating environment (pH, temperature, solvent, auxiliaries, etc.)	Undesired conformation and subsequent leaching result in the reduction of immobilized enzyme activity
Reusability of biocatalysts	Diffusion and mass transfer limitation
Feasibility to be employed in continuous operation	The additional cost of support, preparation materials, and technical inputs
Potential to storage stability	Time-consuming immobilization process

or matrix, allowing them to be employed repeatedly and continuously, lowering the cost of the enzyme applications significantly. Immobilizing biocatalysts can also improve their activity and stability across a wider range of operating conditions, with the added functionality varying depending on both the method of immobilization and the intrinsic qualities of the support materials utilized in such immobilization [11, 46, 87, 94]. Immobilized enzymes are novel in their application due to quick control over reaction, easy separation from the reaction mixture, and simultaneously allowing an appreciable degree of reusability. Table 2 lists some of the most prevalent benefits and drawbacks connected with the usage of immobilized enzymes [11, 50, 87, 92].

Since the early twentieth century, the concept of enzyme immobilization has been explored while by the 1970s, the technology was quite developed and till now an extensive work has been carried out in enzyme immobilization. Today, immobilized enzymes play a critical role in industrial bio-processes, particularly in the food, nutritional, and pharmaceutical industries, as well as in the production of biotechnology products such as diagnostics, bio-affinity chromatography, and biosensors [44, 91, 92]. Immobilization must give high degrees of stability and reusability, as well as additional functionality, without sacrificing enzyme performance or product selectivity, in order to obtain a cost advantage over free enzymes.

7.1 Support Materials

The performance of the immobilized enzyme is strongly influenced by the support materials. Natural and synthetic substrates can be efficiently used as support for immobilization. Biocompatibility, inertness to enzymes, adequate large surface area or porous structure for efficient immobilization with least diffusion limitation in the transport of substrate and product, easy modifying or coating surface to facilitate enzyme immobilization, mechanical stability for reusability, relatively inexpensive, abundant source, and environmentally friendly, etc. are the basic requirements for the support material for immobilization. Also, enzymes should attach to support without changing their conformational and functional properties in order to retain their biological activity [2, 44, 50].

In the early stage of immobilization, support was employed to insolubilize the enzyme and thus facilitate its separation and reuse, which provides easy control over the non-catalytic properties of the obtained immobilized enzyme. Thus, with increased understanding of the correlation of enzyme property with structure and microenvironment, a plethora of synthetic and natural carriers with tailored chemical and physical properties, including various shapes/sizes, porous/non-porous structures, different aquaphilicities, and binding functionalities, have been developed for various bioimmobilization and bioseparation applications [2, 46, 87]. Nanoparticles are now widely regarded as one of the most promising support materials for enzyme immobilization because they provide large surface areas for loading more enzymes, lower substrate mass transfer resistance, reduce protein unfolding, and have unique properties such as mobility, confinement effects, solution behavior, and interfacial properties, all of which significantly improve performance [91, 92].

Low-cost textile matrices, composed of cotton, polyester, or polyamide, are alternative support materials for the immobilization of enzymes. They provide a high enzyme load with high relative reactivity and good durability. Adsorption, covalent bonding, and entrapment can be used to immobilize numerous industrial enzymes onto or inside the textile matrix, resulting in enhanced activity and stability in a multitude of applications [59, 71].

The immobilization of enzymes can be exploited for the design of smart materials in two ways. Enzyme immobilization on textiles can provide materials with new and complex functions, such as antibacterial characteristics and self-cleaning or self-detoxifying qualities [59, 91]. On the other hand, enzymes can be immobilized on smart polymers as support materials. These smart materials respond to triggers (change in pH, temperature, etc.) in various ways, allowing for the regulated release of active compounds such as drugs and perfumes in various applications like controlled drug delivery systems, etc. [92, 99]. Despite the fact that immobilization on supports is a well-established technique, there are no universal criteria for selecting the best support for a certain application.

7.2 Techniques of Immobilization

There are three fundamental techniques of immobilization: carrier binding, entrapment or encapsulation, and cross-linked enzyme aggregates (CLEA). Techniques that provide high levels of stability and reusability, as well as extra functionality without causing substantial hindrances to enzyme activity and product selectivity are important for effective enzyme immobilization. In carrier binding immobilization, the enzyme is linked to a solid support material using the proper techniques to allow for catalytic performance optimization. Physisorption and chemisorption are the two most prevalent techniques for immobilizing carrier binding enzymes. Physical surface adsorption is the most basic and involves low-energy interactions such as hydrogen bonds, van der Waals contacts, and hydrophobic effects to immobilize enzymes on solid supports. Because these physical interactions are insufficient, physisorption is regarded as a reversible form of immobilization in which the enzymes may be easily detached from the support in industrial applications (e.g., high reactant and product concentrations; and high ionic strength). The goal of utilizing an immobilized enzyme on a long-term basis cannot be met with this reversible immobilized enzyme technology [25, 65, 87, 95].

Covalent immobilization has been proven to provide strong chemical bonding, preventing substantial enzyme leaching and mitigating the loss of enzyme active sites. Covalent binding techniques, on the other hand, are more time-consuming and chemically demanding than physical adsorption, and frequently need activation stages capable of causing enzyme denaturation. Furthermore, the choice of an enzyme to be covalently immobilized must be carefully considered in order to achieve optimal catalytic efficiency; for example, the covalent bond between the enzyme and the support must not affect the amino acids associated with the enzyme active site, or the immobilization method may result in catalytic activity loss [2, 11, 44, 70, 95].

The immobilization of enzymes in carriers with various degrees of porosity and permeability is known as enzyme entrapment. Enzymes immobilized in a host support matrix (such as organic or inorganic polymer matrices/hollow fiber membrane/microcapsule) produce a favorable microenvironment that exhibits improved stability and is more catalytically active in organic solvents at higher temperatures, as well as being easily separated from the substrate-product reaction mixture. The use of biocatalysts in the production of organic compounds and new biosensing systems has been investigated using immobilization by entrapment in a range of carriers, such as sol–gels, hydrogels, polymers, and nanomaterials [2, 25, 91, 95].

Enzyme encapsulation in biocompatible nanoparticles and solid supports has also been described as a unique technique for improving enzyme activity as a consequence of biocatalyst–carrier interactions. Encapsulation reduces the issue of leaching or leakage, as well as excessive denaturing [44, 46].

One of the newest classes of immobilization techniques is the formation of cross-linked enzyme aggregates (CLEAs), which are carrier-less macroparticles formed by a covalent bond involving amino acid residues ($-NH_2$, $-CO_2$, $-SH$) of the enzyme by cross-linking with another to form cross-linked enzyme crystals (CLECs) and cross-linked enzyme aggregates (CLEAs). The usual production of CLEAs involves aggregating supplied soluble enzymes with a precipitating reagent, such as ammonium sulfate, acetone, ethanol, or tert-butanol, and then copolymerizing the enzyme aggregates with a cross-linking agent, most commonly glutaraldehyde [40, 50, 94].

The research revealed, however, that aggregate cross-linking is not a universal immobilization technique, and that it should be tailored to each target biocatalyst, with the precipitating and cross-linking agents carefully chosen to guarantee that immobilization does not impair enzyme performance. In other cases, CLEAs were shown to have greater catalytic activity than their free counterparts, and this phenomenon, known as hyperactivation, was ascribed to the enzyme aggregating in a preorganized tertiary structure that rendered it irreversibly insoluble following cross-linking. As a result of their high catalytic productivity and low-cost immobilization techniques, CLEAs have a lot of promise for use in industrial processes. However, enhancing CLEA's mechanical characteristics and better specifying separation criteria for continuous operations were required for successful application scale-up [44, 94, 95].

Techniques such as covalent binding, cross-linking enzyme aggregation, and entrapment or encapsulation, among others, are categorized as irreversible enzyme immobilization, which includes strong chemical interactions and is used to keep enzymes stable for a long period. The majority of industrial applications use enzymes immobilized via these approaches, allowing for continuous processing without the need to replenish the enzyme on a regular basis [40, 46]. Since then, several articles have reviewed and published techniques and tactics for enzyme immobilization.

7.3 Performance of Immobilized Enzymes

Immobilization has a significant impact on enzyme characteristics including activity, stability, pH and temperature optimum, kinetic parameters, and substrate selectivity [70, 87]. The performance of immobilization can be assessed in terms of catalytic and non-catalytic activities of immobilized enzymes. The catalytic functions of enzymes that convert substrates into products are linked to activity, selectivity, and stability, whereas non-catalytic functions include the physical properties (like mass transfer, recovery, and reusability) of immobilized enzymes and are strongly related to the physical (shape, size) and chemical nature of the non-catalytic part, i.e., support [12, 70].

The catalytic activity of a native enzyme is primarily dependent on its intrinsic structure and operational conditions, whereas the microenvironment plays a crucial role for an immobilized enzyme that includes carrier's physical (shape, size, and thickness) and chemical structure, the nature of interaction between enzyme and support, binding position and number of bonds, properties of the spacer used, and the support's conformational freedom [12, 25, 70].

One of the objectives for immobilizing enzymes is to enhance enzyme stability during long-term storage or under various working conditions, such as changing temperatures or pH values. Researchers generally compare immobilized and native enzymes to measure these characteristics. On immobilization, the stability of the immobilized enzymes may increase or decrease depending on time, temperature, other storage conditions, and experimental factors [56]. Quite frequently, operational stabilization is noticed as a result of loading an excessive amount of enzyme. True molecule stability, on the other hand, might be due to the multipoint attachment. Although high enzyme loading during immobilization is expected to promote activity, limited diffusion reduces perceived activity [54, 57, 95]. Excessive enzyme loading always results in protein–protein interaction, which prevents the flexible stretching of enzyme structure, resulting in steric hindrance and, as a result, enzyme deactivation. Under molecular crowding circumstances, the enzyme molecule may find it difficult to modify its most appropriate shape for catching substrate molecules and releasing product molecules [54, 56, 57].

It is critical to assess the immobilization technique yield, which is calculated by combining the proportion of enzyme immobilized and the residual enzyme activity after immobilization under the experimental circumstances. The percentage of enzyme immobilized is generally expressed in terms of bound protein, which is determined by subtracting the quantity of enzyme left in the supernatant after immobilization from the initial amount [25, 48].

It is more difficult to identify the absolute enzyme activity that remains on the support after immobilization; therefore an apparent or specific activity is generally assessed instead, which accounts for mass transfer and diffusional limitations in the experimental technique. The fact that an immobilized enzyme is attached to support and has purposefully restricted mobility causes this effect. In addition to diffusional limitations, the support's charge might have partitioning effects; therefore, it is vital to examine immobilized enzyme activity under a range of ionic circumstances when defining them [25, 48, 65].

The effectiveness factor, which determines the effectiveness/efficiency of an immobilized enzyme with a native enzyme under proper reaction circumstances, is generally represented as the reaction rate/unit mass of the protein generated. The Michael's constant Km and maximal reaction velocity V_{max} are two kinetic parameters that are frequently computed for an immobilized enzyme to examine the impact of immobilization on the enzyme's catalytic efficiency when compared to the non-immobilized enzyme counterpart. Km compares the rates of substrate-enzyme binding and dissociation, with lower Km values indicating that binding takes precedence and indicating stronger enzyme_substrate affinity. V_{max} is a good measure of catalytic activity because it evaluates how quickly an enzyme converts a substrate to a product when it is controlled for catalytic mass [15, 54, 56].

Operational and storage stability are the other key performance metrics. Storage stability is defined as monitoring activity after a certain period of time under specified storage circumstances, whereas operational stability is defined as activity degradation under working settings such as pH, temperature, solvent, and so on. These metrics give information on performance, which may be represented in terms of productivity (product generated per unit enzyme) or enzyme consumption (enzyme consumed per unit product produced). The pH and temperature optima (for operating range), the apparent Michael's constants (Km) for suitable substrates (which also show immobilized enzyme selectivity), and lastly the reusability cycles are all measurable characteristics for an immobilized enzyme [48, 56]. The possibility for recycling and reuse of enzymes as biocatalysts inspired the main notion of immobilization. To offset the expenses of enzyme immobilization, it is very desirable that the immobilized enzyme may be reused for several cycles.

The immobilization variables described above will have a major impact on the immobilized enzyme's properties. True immobilized enzymes should have a well-balanced overall performance, with adequate immobilization yields, low mass transfer constraints, and good operational stability [48, 56]. Thus, along with the selection of immobilization supports, conditions, and methods, the overall enzymatic

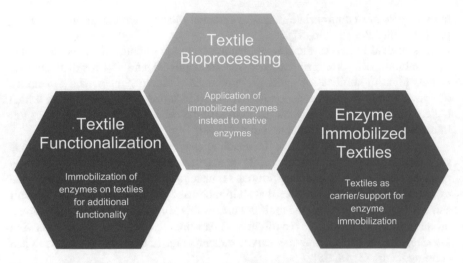

Fig. 4 Immobilization as sustainable solutions in textile chemical processing

activity, enzyme deactivation characteristics, cost of immobilization procedure, toxicity of immobilization reagents, and desired final properties of immobilized enzymes should all be taken into account when developing immobilized biocatalysts [12, 56].

8 Immobilization of Enzymes as Sustainable Solutions for Textile Processing

Immobilization of enzymes has three different perspectives in textile chemical processing (Fig. 4):

- Application of immobilized enzymes in bio-processing of textiles
- Immobilization of enzymes on textiles for functionalization
- Enzyme immobilized textiles

9 Application of Immobilized Enzymes in Bio-processing of Textiles

In conventional enzymatic processing, the enzymes are drained after a single use which is not an economical practice as enzymes are costly. Immobilization refers to the attachment of enzymes to a solid structure that restricts movement completely or to a small limited region. The primary advantage of enzyme immobilization is that it is simple to remove the enzyme from the reaction mixture (substrates and products) and that it may be reused several times, resulting in considerable cost savings for the enzyme and enzymatic products. In addition to that immobilization allows the quick start and stops of the reaction by moving the enzyme into and away from the reaction solution and increases the stability of enzymes by monitoring reaction conditions. Immobilization also restricts the diffusion of enzymes in the fiber substrate and reduces the probability of fiber damage in various enzyme applications. Thus, immobilization adds more sustainability to the eco-friendly approach of enzymes application in textile processing [2, 92, 95].

9.1 Bio-desizing Using Immobilized Amylases

Bio-desizing is considered an environmentally benign alternative to traditional desizing in the modern textile industry since it produces less effluents and contains less harmful chemicals. The α-amylase enzyme acts at random locations along the amylose and amylopectin chains, hydrolyzing starch into water-soluble oligosaccharides that can be readily washed away later. Salts, wetting agents, non-ionic surfactants, and ultrasound can assist improve enzyme efficiency by enhancing fiber swelling or promoting enzyme penetration. The assistance of lipase with amylase not only helps in starch removal due to synergism but also helps in removing the other natural impurities and shortens the processing time. The optimal temperature and pH for each amylase are different, but in general, the best temperature is between 55 °C and 70 °C, while the optimum pH is between 5.0 and 7.5. Protein engineering is used to produce modified amylases with enhanced performance. Amylases from thermophiles may work in a wide range of pH and temperature conditions, making them viable green alternatives for commercial use. Even though enzymatic desizing is commercial at an industrial scale owing to inexpensive and industrially viable amylases, immobilization can add easy separation of amylase from size residue, resulting in not only enzyme reusability but also size recovery. This can overcome the primary disadvantage of traditional enzymatic desizing wastewater, which is its high biological oxygen demand (BOD) [18, 50, 85].

The use of ultrasonic energy to desize grey cotton fabric with immobilized amylase was described by Sahinbaskan in 2011. Even after five cycles, immobilized amylase on acrylated epoxidized soyabean oil (AESO) resin retains 50% of its initial activity, allowing it to be recovered and reused. Furthermore, sonification enhances starch removal with immobilized enzyme while reducing strength loss when compared to free amylase [81]. Immobilized commercial α-amylase onto chitosan and Eudragit S-100 can be easily separated by reversible soluble–insoluble polymer supports and can be reused up to four repeated cycles. Eudragit S-100 is found to be a more suitable support material than chitosan for the desizing application of α-amylase [51].

9.2　Bio-scouring Using Immobilized Pectinases

Bio-scouring is a low-energy, environmentally acceptable alternative based on the notion of using enzymes to specifically target non-cellulosic contaminants. Enzymatic scouring allows you to efficiently scour cloth while causing no harm to the fabric or the environment. When compared to alkaline scouring, the enzymatic scouring method has a 20–45% lower biological oxygen demand (BOD) and chemical oxygen demand (COD) (100%). Bio-scouring formulations including pectinase and lipase are more successful at achieving good cellulosic textile hydrophilicity. Cellulases, on the other hand, enhance impurity removal by hydrolyzing cellulose chains. For this purpose, mixed enzyme systems containing pectinase, lipases, proteases, cellulases, xylanases, or cutinases might be highly efficient. Bio-scouring has yet to be commercialized on a large basis, despite much study. There is a demand for a cost-effective pectinase with greater activity and resistance to high temperatures and alkalinity [18, 50, 58, 86].

Pectinases have a lot of promise for being used efficiently if immobilization principles are applied to them. Until now, pectinases have been immobilized on a variety of supports, including polyacrylonitrile copolymer membranes, ion exchange resins, bone as a solid support, aminated silica gel, cross-linked enzyme crystals, and cross-linked enzyme aggregates, macroporous polyacrylamide, polygalacturonase on activated polyethylene, polymer nanocomposite microspheres, silica-coated chito used a pectinase immobilized in a reverse micellar system to successfully expose cotton to scouring. Scouring's efficiency was comparable to or better than that of traditional alkaline procedures or bio-scouring in aqueous environments. Even in organic media, the enzyme showed outstanding activity [50, 92].

9.3　Bio-bleaching Using Immobilized Glucose Oxidase

Glucose-oxidase enzymes are electron-transfer oxidases that catalyze the conversion of glucose to glucolactone in the presence of oxygen, yielding gluconic acid and H_2O_2 as a side product for bleaching. Peroxide generation using glucose oxidase requires slightly acidic to neutral conditions at low temperatures, whereas the bleaching for bleaching, a high temperature of 80°–90 °C, and an alkaline pH of 11 produce effective results. To enable bleaching at low temperature in neutral media, several researchers investigated the activation of generated hydrogen peroxide with different activators such as peroxidase enzymes, and peracid precursors such as TAED (tetraacetylethylenediamine), NOBS (non-anoyloxybenzene sulfonate), and TBCC (N-[4-(triethylammoniomethyl) benzoyl] caprolactam chloride). This enzymatic system in cotton pretreatment is especially advantageous since desizing waste baths may be reused as a glucose source, reducing wastewater pollution and water usage. Desizing and bleaching in one bath or bleaching with a reused desizing bath have both been investigated in several researches [50, 52, 97].

Researchers discovered that a higher concentration of enzymatically produced peroxide was needed to attain the same whiteness as traditional H_2O_2. In fact, the presence of glucose and enzyme in the in situ created H_2O_2 bleach solution causes H_2O_2 to become over-stabilized, resulting in decreased whiteness. Also raising the bleaching temperature leads the enzyme protein to denature and deposit on the fabric surface owing to hydrophobic fabric/enzyme interactions. As a result, immobilization allows the enzyme to be separated from the bleach solution, resolving these difficulties and increasing whiteness. Furthermore, immobilization permits the costly glucose-oxidase enzyme to be reused, saving money [50, 52, 97].

According to multiple researches, immobilization techniques on various supports improved the stability of glucose oxidase. In first textile use, with cyanuric chloride-mediated attachment to cotton and subsequent release of H_2O_2 at quantities sufficient to permit effective bleaching. Glucose oxidase was covalently immobilized on two affordable and commercially accessible supports, namely alumina and glass, in another study. Due to its greater porous shape, glass generated more H_2O_2 (0.35 g/L) and hence bound more protein; nevertheless, alumina created superior operational stability and enabled reuse at three consecutive cycles [97]. In another study, it was discovered that roughly 30% more H_2O_2 (in situ created) provides whiteness similar to traditional H_2O_2, but glucose-oxidase immobilization on chitosan can lower this amount to 10%. The immobilization permits the glucose-oxidase enzyme to be recovered, and proper bleaching was achieved in three consecutive applications. Therefore, it can be reused three times for hydrogen peroxide generation instead of free which is generally drained after single usage [52].

9.4 Bio-polishing Using Immobilized Cellulase

Bio-polishing is the process of employing cellulases to modify the surface of cellulosic textiles to give a softer feel, a more luminous color, and better resistance to pilling. The removal of superficial microfibrils of cotton fibers by cellulase enzymes is achieved by controlled partial hydrolysis of cellulose followed by mechanical treatment; the result is a softer finish and a uniform surface. Acid cellulases enriched with endoglucanase are highly suited for the bio-polishing of cellulosic fabrics [18, 50, 86].

The free cellulase used for bio-polishing may easily permeate through the cotton's peripheral cellulose structure, resulting in significant weight and strength loss, which is not the goal of bio-polishing. Thus, immobilization of cellulase, or attachment to a carrier, might limit the enzyme's activity to the fiber surface and avoid losses [45]. Several approaches and supports for immobilizing cellulases for diverse purposes have been disclosed too far, all of which improve thermal stability and reusability. Acid cellulases are more often used for bio-polishing and denim fading since they are less expensive. However, their use causes more abrasion, which reduces the tensile strength of cotton fibers. As a result, particularly for textile applications, Dincer and his co-worker immobilized an acidic cellulase on chitosan beads covered with maleic anhydride-modified polyvinyl alcohol to increase its stability in the neutral pH range. Immobilization shifts the enzyme's pH optimum from 4 to 7, converting acid cellulase to neutral cellulase. Cellulase immobilized on chitosan beads outperformed native enzymes in terms of pH stability [27].

Similarly, to retain the cellulase after bio-polishing, commercial liquid cellulase (Biowash-L) was immobilized on ion exchange (fermer) and epoxy resin (Ceralite IRC50) carriers using methanol. Immobilization on epoxy resin exhibits a higher level of cellulase immobilization and exceptional durability throughout consecutive application cycles as compared to ion exchange resin supports. The bio-polishing effect on cotton fabric with immobilized cellulase was good for six successive cycles with less weight and strength loss compared to free enzyme. Thus it is very clear that cellulase enzyme can be effectively conserved by immobilization techniques and is quite capable of being reused for several numbers of cycles [45]. Cellulase was successfully immobilized on different smart (soluble–insoluble reversible with pH change) polymers Eudragit L-100 and N-succinyl-chitosan in many kinds of literature and could be the best alternative to apply in bio-polishing and/or bio-scouring [28].

9.5   Denim Bio-wash Using Immobilized Cellulase

Denim is popular cotton casual because of its unique faded and worn look. The cellulase enzyme causes non-homogeneous surface removal of the indigo dye contained inside the fibers, giving it an aged look as well as improved softness and flexibility. Denim finishing was the first application of cellulases in textile processing in the late 1980s. Bio-washing with cellulase enzyme is now a viable option for achieving a desirable look and excellent quality while also being environmentally friendly [18, 29]. The biggest issue is back-staining or the re-deposition of released indigo onto the white section of denim clothes after denim washing. Back-staining is caused by the high affinity of indigo and cellulase enzymes, as well as the strong binding of cellulases to cotton cellulose. Furthermore, the application of cellulases enzyme is not limited to the surface of the material, resulting in a significant loss of strength. As a result, immobilization may be a viable option for overcoming these issues, with the added benefit of being able to extract the enzyme for future use [32, 58].

A commercial cellulase was immobilized onto ZrOCl2-activated pumice particles for denim washing. In comparison to commercial acid cellulase and simple pumice washing, the immobilized enzymes were able to efficiently abrade indigo-dyed denim fabrics and generate higher L values (CIE). A similar fading effect was also found in immobilized cellulase from the first to the fifth cycle. As the strongly bound enzyme persists on pumice support, staining was found at lower levels after the second usage with immobilized cellulase than its native counterpart [75].

In another study, native and cellulase immobilized on Eudragit S-100 were used to study the enzymatic treatment of denim fabric. Immobilized cellulase may effectively remove indigo dyestuffs from denim textiles' surfaces without causing considerable weight loss or mechanical damage. Covalently immobilized Eudragit–cellulase demonstrated decoloration and a color effect similar to native cellulose. Therefore, the immobilized cellulase has great potential in the bio-washing of denim fabrics without the problem of excessive damage to fibers [102].

9.6 Shrink-Proofing of Wool Using Immobilized Enzymes

Protease treatment for wool shrink resistance is a more environmentally friendly option than regular chlorine treatment. The protease treatment, on the other hand, produces severe damage to the fiber cuticle, resulting in decreased fiber strength and shrink resistance. Increasing enzyme size by chemical cross-linking with glutaraldehyde or the addition of synthetic polymers such as polyethylene glycol has been shown in several studies to minimize enzyme penetration and, as a result, strength and weight loss. Immobilization of proteases, on the other hand, often increases their molecular size; the modified protease stayed on the surface of the cuticle layer region (hydrolyzing just the cuticle layer of wool), resulting in less tensile strength loss and optimal fiber felting [83, 88, 89].

Carbodiimide coupling was utilized to covalently attach a commercial protease (Esperase) to Eudragit S-100 for wool shrink-resist finishing (soluble–insoluble polymer). The immobilized enzyme exhibit low specific activity with higher thermal stability while the optimum pH shifts toward a slight alkaline side. The wool fabric treated with immobilized Esperase exhibited higher shrink resistance and reduced fiber damage in terms of weight loss and tensile strength when compared to native Esperase. This innovative approach presents a possible alternative for wool shrink-resist treatment since the proteolytic assault is limited to the cuticle surfaces of wool fibers due to enzyme immobilization [88, 90].

9.7 Bleach Clean-up Using Immobilized Catalases

Catalases are enzymes that catalyze the breakdown of H_2O_2 into water and oxygen, making them the best option for bleach bath clean-up before reactive dyeing [4, 21]. Several authors have tried to immobilize catalase using organic and inorganic materials such porous glass, cellulose, alumina, silica gel, and hydrogels. Some natural polymers, such as gelatin and chitosan, and some synthetic polymers, such as polyacrylamide, have been discovered to be unsuitable for bleaching procedures. Opwis in a series of research work exploited synthetic textiles as support material for covalent immobilization of catalase using photochemical process which is not only economical but also easy to remove without filtration [20, 58].

Costa and his co-workers used alumina-based supports that are also suitable for catalase immobilization as their superior mechanical stability at high pH and temperatures. The bleached solution treated with immobilized catalase (Bacillus SF) was utilized for dyeing textiles with various colors, resulting in acceptable color variations for all colors employed [20]. For the recycling of textile bleaching effluents for dyeing, a commercial catalase (Terminox Ultra 50 L) was covalently immobilized on alumina using glutaraldehyde [21].

9.8   Polymerization Using Immobilized Enzymes

Enzymes, especially the immobilized enzymes, have significant potential in the polymerization step of fiber manufacturing, as well as the most common immobilization techniques used in this field [56, 65]. Among the enzymes used efficiently for polymer synthesis, Candida antarctica lipase B (CALB) is by far the most well-known in the literature. Chemical reactions such as transesterification, Michael addition, and ring-opening polymerization may be carried out at lower temperatures using CALB, because of CALB's catalytic activity. It is employed as an immobilized enzyme on silica material in the ring-opening polymerization of caprolactum to polycaprolactone in the majority of reported enzymatic polymerizations [31]. Novozym 435, a physically immobilized form of CALB in a macroporous polymethyl methacrylate resin, is commercially available. Graphite-immobilized glucose oxidase has been proposed as a viable catalyst for the oxidative polymerization of polythiophene at neutral pH under ambient settings in other studies [43].

9.9   Detergents Containing Immobilized Enzymes

Immobilized enzymes can be used to make synthetic detergents that remove dirt and stains more effectively. Proteases, amylases, cellulases, and other enzymes are immobilized with detergent granulations. These enzymes are relatively stable and active even when subjected to severe washing conditions. Lipases are enzymes that break down fats and oils to eliminate stains. Proteases remove protein stains such as those caused by blood, egg, and sweat. Amylases break down carbs like chocolate and gravies to remove stains [15, 30, 92].

Esperase immobilized on Eudragit S-100 can be used in various detergent formulations. Human blood and egg yolk stains were removed from cotton and wool fabric samples using immobilized Esperase, which was shown to be more stable than the native enzyme. As a result, they can be used as a detergent component to improve washing results [98]. With roughly 75% immobilization, both the agarose and agar matrixes showed to offer effective support for the enzyme at a low cost. Immobilized amylase can help detergents work better by removing starch stains from textiles, hence it might be used as a detergent ingredient [76]. Furthermore, by glutaraldehyde-mediated coupling, an amylase from Aspergillus niger was immobilized onto zirconia-coated alkylamine glass beads and used with various detergents to remove starch stains from cotton clothing. All detergents functioned better in the presence of immobilized amylase.

9.10   Dye Synthesis Using Immobilized Enzymes

Laccases and peroxidases enzymes may catalyze the oxidation of phenolic and aromatic compounds to produce colorful molecules that can be utilized as dyes, which is very beneficial in the textile industry. To synthesize new compounds from a range of aromatic substrates, they utilize a variety of dimerization, oligomerization, and polymerization techniques. Laccase oxidizes dye precursors such as phenols, aminophenols, and diamines to aryloxy radicals, which can subsequently undergo non-enzymatic reactions to produce colorful compounds. Fungal laccases, in particular, have a high oxidative potential, allowing them to be employed as an oxidative catalyst in the textile industry's de novo color creation. Laccase catalyzes the oxidation of single benzene derivatives with at least two substituents (amino, hydroxy, and methoxy groups) to yield brightly colored products (varying from yellow/brown to red and blue to green). Laccase-based textile coloring patents list hydroxy, methoxy, methyl, methylene, sulfonic, carboxy derivatives of benzene, naphthalene, anthraquinones, and heterocyclic precursors as precursors and couplers. Furthermore, laccase-mediated procedures for the manufacture of azo- and phenoxazinone-type colorants are only mentioned in a few patents. Laccase-mediated fiber coloring, as detailed in Novozymes patents, for cotton, linen, cellulose, flax, polyacrylic, polyamide, polyester, viscose, and other fibers [77, 78].

Laccase, a phenoxazinone chromophore present in actinomycins, was isolated from Trametes Versicolor and used to convert 4-methyl-3-hydroxyanthranilic acid to 2-amino-4,6-dimethyl-3 phenoxazinone-1,9-carboxylic acid (actinocin) [72]. Similarly, for color generation, immobilized white-rot fungus strains were utilized as less expensive industrial-grade biocatalysts. The ability of such biomass to catalyze the transformation of benzene and naphthalene derivatives into stable and non-toxic polymers with good dyeing properties highlights the possibility of transforming precursors into dyes more efficiently and simply using fungal biomass rather than isolated enzymes [78]. According to a research, the homocoupling reaction of 4-methylamino benzoic acid on Eupergit C, which was mediated by immobilized Trametes Versicolor laccase, produced colorful azo intermediates [53].

The pigment was synthesized using LbL immobilized tyrosinase and laccase, with the lignin scaffold serving as the pigment's carrier and binder. To create a vast panel of colors, L-DOPA, L-tyrosine, epicatechin, and other phenol derivatives were utilized as reactants. Pigmentation processes are catalyzed by reactive lignin nanocapsules, resulting in a unique sort of sustainable polyvalent bio-ink. Bio-inks in a variety of colors, including black, grey, yellow-like, pink-like, and red/brown hues, have been created using a combination of reactants and pigmentation techniques [13].

To make indigo and indigo-derivatives from indoles, a peroxygenase method catalyzed by the CYPBM3F87A Bacillus megaterium immobilized on magnetic nanoparticles is used. This biocatalyst is easy to employ, and its magnetic properties will make it easier to recover in a large-scale operation. It looks to be a viable ecologically friendly choice for industrial dye manufacturing [55].

The laccase-mediated transformation of 2-amino-3-methoxybenzoic acid into the novel and environmentally friendly orange dye (N15) was produced using native and immobilized laccase (LAC) from Pleurotus ostreatus strain, and the dye can effectively color wool, cotton, silk, and flax fibers at concentrations of 0.2 and 0.5% [100].

10 Textile Wastewater Treatment Using Immobilized Enzymes

Sustainability in the world has raised the concept of socioeconomic growth with an emphasis on pollution reduction. Enzymes catalyze the oxidation of pollutants to simpler compounds with lower toxicity that can be readily separated from the post-reaction mixture during biodegradation. Laccases, peroxidases, and tyrosinases are examples of enzymes from the oxidoreductase family that have emerged as a viable alternative for biological wastewater treatment [22, 46, 58, 92].

The usage of enzymes in their free form has certain drawbacks, such as low stability and recovery difficulties, which restrict their reusability. Immobilized enzymes were proposed as flexible biocatalysts for wastewater degradation as immobilization enhances the temperature and pH stability of the enzymes, allowing them to be reused in further bioremediation cycles. Furthermore, the use of immobilized enzymes offers the possibility of novel bio-removal strategies such as simultaneous adsorption and biodegradation or the use of enzymatic membrane bioreactors, thus it may become a feasible alternative for industrial application [46, 93, 103].

Synthetic dyes are currently one of the most hazardous categories of pollutants in the environment. This is owing to a rise in the usage of anthraquinone, azo, and triaryl-methane dyes in textiles and other everyday products. The bulk of dyes are released as wastes directly from the textile industry following dying operations; nevertheless, these chemicals easily enter into household wastewater as laundry wastes. Due to inadequate wastewater treatment, dyes end up in various water bodies, including seas, rivers, and even groundwater. The dye may easily accumulate throughout the food chain, and its toxicity and carcinogenic properties might disrupt the physiological systems of the ecosystem [47].

Therefore, immobilized enzymes are essential for dye removal from wastewaters and decrease solution toxicity after treatment. Previous research has confirmed that oxidoreductases are effective in the degradation of dyes [46]. It is worth noting that the usage of free and immobilized oxidoreductases often results in over 90% pollutant elimination. Frequently utilized enzymes for wastewater treatment applications are laccases and peroxidases. The immobilized form of these oxidoreductases has recently been reported to be utilized to decolorize dyes from real textile wastewaters.

Peroxidases may be a useful tool for not only decolorizing dye solutions but also for reducing dye toxicity after enzymatic treatment [14]. Table 3 shows a thorough overview of the use of immobilized enzymes in textile wastewater treatment [46, 60, 103].

Apart from the kind of immobilized enzymes, the support material has a major impact on the dye degradation process. However, to achieve highly active biocatalytic systems, the support material for immobilized enzymes must be selected carefully. For oxidoreductase immobilization, organic, inorganic, and hybrid/composite materials are widely employed. As enzyme supports, natural and synthesized polymers are appealing. Their multifunctional application for enzyme immobilization and wastewater treatment is enabled by the functionalization of their surface. Although many materials can be employed, hybrid and composite materials are of special interest due to their tailor-made characteristics that assure high enzyme activity and stability [1, 42, 60].

The immobilization of these enzymes is mostly focused on silica and its composites, which provide reliable and cost-effective support. Another significant category of chemicals utilized as laccase carriers is functionalized Fe_3O_4-based materials. Because the generated support contained magnetic particles, it was possible to separate the biosystem from the dye solution following treatment using a simple external magnetic field [3]. Titanium dioxide-based photocatalysis is advantageous due to its inexpensive cost, lack of toxicity, strong photocatalytic activity, and long-term stability. Another biopolymer widely utilized for enzyme immobilization is chitosan [49]. Incorporating some inorganic elements, such as clays, improves the practical operability and effectiveness of chitosan immobilization [7, 46, 60, 103].

The efficacy of enzymatic decolorization of colors from textile effluents is influenced by enzyme origin, support type, the presence of additional chemicals in solutions such as salt ions and surfactants, and process parameters. The immobilization of oxidoreductase in wastewater treatment has a lot of promise. However, in order for laboratory-scale benefits to be translated to higher scales for real-world wastewater treatment, a number of obstacles must be overcome swiftly. More study is needed on immobilization and bioremediation settings, as well as the use of immobilized biocatalysts for continuous treatment of model and real water solutions [14, 60, 79, 103].

Table 3 Summary of application of immobilized enzymes in textile wastewater treatment

Enzymes and Source	Support material	Immobilization technique	Pollutant tested	Effectiveness (degradation or decolorization % and reusability)
Laccase enzymes				
Laccase	Poly(p-phenylenediamine) /Fe$_3$O$_4$ nanocomposite	Covalent binding	Reactive blue 19	90%
Laccase	Silica modified with 3-glycidyloxypropyltrimethoxysilane (3-GPTMS)	Covalent binding	Acid orange 156 Acid red 52 Coomassie brilliant Blue Methyl violet Malachite green	40–50%
Laccase from *Trametes versicolor*	Amino-functionalized magnetic Fe$_3$O$_4$ nanoparticles	Covalent Co-immobilization with 2,2,6,6-tetramethylpiperidine-1-oxyl (TEMPO nanoparticles)	Acid Fuchsin	77% 50% residual activity after 8 repeated cycles
Laccase	Titania modified with epoxy-containing polymer	Covalent binding	Rhodamine B	95%
Laccase from *Biotitania*	Titanium (IV) bis (ammonium lactato) dihydroxide (Ti-BALDH/TiO$_2$	Covalent binding	Malachite green	90%
Laccase from *Aspergillus species*	Porous polyvinyl alcohol/ Halloysite hybrid beads	Covalent binding	Reactive blue	50%, More than 60% in 6th cycle

(continued)

Table 3 (continued)

Enzymes and Source	Support material	Immobilization technique	Pollutant tested	Effectiveness (degradation or decolorization % and reusability)
Laccase from *Trametes villosa*	Thiolsulfinate-agarose (TSI-agarose)	Covalent reversible	Acid red 88 Acid black 172	77% and 70% decolorization level of 78% in the third use
Laccase *Bacillus* sp. MSK-01	Cu-alginate bead	Entrapment	Textile dye effluent	100% even after 4 cycles
Laccase from *Escherichia coli*	Poly-hydroxybutyrate (PHB) beads	Covalent binding	Direct red 105, direct yellow 106, direct black 112	N.A
Commercial Laccase	Oxidized activated carbon	Covalent binding	Reactive blue 19	N.A
Laccase from *Escherichia coli*	Magnetic zeolitic Imidazolate framework-8 (Fe_3O_4@ZIF-8)	Covalent binding	Indigo carmine	Completely decolorize after five consecutive cycles
Laccase	Fe_3O_4@C-Cu^{2+} nanoparticles	Covalent binding	Malachite green Brilliant green Crystal violet Azophloxine Procion red MX-5B reactive blue 19	99% 93% 79% 88% 75% 81% More than 65% removal efficiency after 10 reuses

(continued)

Table 3 (continued)

Enzymes and Source	Support material	Immobilization technique	Pollutant tested	Effectiveness (degradation or decolorization % and reusability)
Fungal laccase from *Pycnoporus sanguineus* CS43	Silica nanoparticles	Covalent binding	Congo RED	Textile filter-based bioreactors give 39% efficiency
Laccase	Magnetic metal-organic frameworks (MMOFs)	Covalent binding	Crystal violet Methylene blue	98% 96% Continuous setup
Laccase nanoflowers	Fe_3O_4 magnetic nanoparticles	Encapsulation	Malachite green	90% degradation efficiency after 18 reuses
Laccase	Iron oxide modified with thiolated chitosan (TCS)	Covalent binding	Reactive blue 171 Acid blue 74	N.A
Laccase from *Alcaligenes faecalis XFI*	Chitosan beads (CB) Chitosan-clay composite beads (CCB)	Covalent binding	Methyl red Remazol brilliant blue reactive black 5	85% 85% 69%
Laccase from *Trametes versicolor*	Poly(methyl methacrylate)/Polyaniline electrospun fibers	Covalent binding	Remazol brilliant blue	60%
Laccase	Polydopaminecoated Poly(vinylidene fluoride) membrane (PDA@PVDF)	Covalent binding	Congo red	90%, 80% after 5 repeated cycles

(continued)

Table 3 (continued)

Enzymes and Source	Support material	Immobilization technique	Pollutant tested	Effectiveness (degradation or decolorization % and reusability)
Co-immobilized spore laccase/TiO$_2$ nanoparticles	Alginate beads	Entrapment	Alizarin red Trypan blue Malachite green Indigo carmine	60% 96% 100% 100%
Laccase from *Trametes versicolor*	Magnetic amino-functionalized metal–organic framework Fe$_3$O$_4$-NH$_2$@MIL-101(Cr)	Covalent binding	Reactive black 5 Alizarin red S	55% 91% Five reuse cycles
Laccase from *Trametes vesicolor*	Glycidyl methacrylate (GMA) functionalized Polyacrylamide-alginate cryogel (PAG)	Covalent binding	Textile wastewater	50% after 5 cycles
Laccase from *Pleurotus nebrodensis* WC 850	Glutaraldehyde cross-linked chitosan beads	Covalent binding	Reactive and disperse dyes	(83–90%) 54% after 8 cycles
Laccase from *Streptomyces cyaneus*	Dopamine-modified pectin	Entrapment	Amido black 10B Reactive black 5 Evans blue	60% of azo dyes after 10 cycles
Peroxidase Enzymes				
Manganese peroxidase from *Phanerochaete chrysosporium*	Chitosan beads	Cross-linking	Mixed textile effluent	60% after 10 cycles
Horseradish peroxidase (HRP)	Silica coated with zinc oxide	Covalent binding	Anthraquinone dyes	100% decolorization

(continued)

Table 3 (continued)

Enzymes and Source	Support material	Immobilization technique	Pollutant tested	Effectiveness (degradation or decolorization % and reusability)
Chloroperoxidase (CPO) and Horseradish peroxidase (HRP)	ZnO nanowires/macroporous SiO$_2$	Cross-linking	Azo dyes	Complete decolorization of dyes within 3 h
Chloroperoxidase (CPO) and Horseradish peroxidase (HRP)	Polydopamine-tethered CPO/HRP-TiO$_2$ nanocomposite	Covalent binding	Aniline blue	100% removal
Peroxidase *Pseudomonas aeruginosa* strain OS4	Fe$_3$O$_4$ magnetic nanoparticles modified with glutaraldehyde	Covalent binding	Direct green Reactive blue	Completely decolorized in 6 h
Horseradish peroxidase (HRP)	Graphene oxide–silica modified with 3-aminopropyltriethoxysilane	Covalent binding	Colored effluent	100% decolorization of dyes
Horseradish peroxidase (HRP)	Purolite® A109	Covalent binding	Anthraquinone-based synthetic dye	90%
Lignin peroxidase *Pseudomonas fluorescens* LiP-RL5	Graphene oxide functionalized MnFe$_2$O$_4$	Covalent binding	Methylene blue	88%
Manganese peroxidase from *Anthracophyllum discolor*	Fe$_3$O$_4$/Chitosan	Covalent binding	Methylene blue Reactive orange 16	86% after 5 cycles
Horseradish peroxidase (HRP)	Polyamide 6	Adsorption/covalent binding	Reactive black 5 Malachite green	70% after 20 cycles
Peroxidase from *Moringa oleifera Lam* (MoPOX)	Glutaraldehyde-activated chitosan beads	Covalent binding	Textile wastewater	More than 80% removal
Glucose oxidase	Polyacrylonitrile (PAN)-based carbon felt	Covalent binding	Remazol® Blue RR	93%

11 Immobilization of Enzymes for Textile Functionalization

Different enzymes have been immobilized on various textile surfaces, including cotton, polyester wool, and flax, to achieve additional functionalities. Several naturally occurring enzymes, namely hydrolase enzymes, have recently been discovered as possible biological protective finishes, offering antibacterial activity as well as a potential barrier to microbial entry via cell wall polysaccharide hydrolysis. One of these is lysozyme, which has been shown to be effective against a wide range of bacteria. Its immobilization on textiles has been used to create smart materials that have unique properties including self-cleaning and antibacterial properties [16, 74].

Lysozyme, for example, has been immobilized on cotton fabrics to improve antibacterial properties and prevent microbial damage. Researchers have evaluated the antibacterial efficiency of lysozyme attached to glycine derivatized cotton textiles (non-woven and woven) for potential usage in non-woven wipes, clothing, and protective fabrics [16, 35].

Immobilization lysozyme with glutaraldehyde and microbial transglutaminase as cross-linkers has been used to make wool-based antibacterial textiles. The cloth with covalently attached lysozyme was rather durable, and the activity was preserved after numerous washing cycles [84]. When compared to natural lactoferrin, wool fabric immobilized with lactoferrin employing microbial transglutaminase as a biological cross-linker has improved antibacterial activities against *Staphylococcus aureus* and *Escherichia coli*. This enzymatic approach of wool activation gives an antibacterial finishing treatment that is safe for the environment [89].

Additionally, sodium periodate was employed to covalently immobilize bovine pancreatic trypsin on sterilized cotton gauge bandages for use as an anti-inflammatory wound dressing. Immobilized trypsin has better operational and storage stability since no trypsin is released after 30 days [84]. Peroxidase and laccase increase the technical properties of textiles by connecting functional molecules. Laccase is used to graft lauryl gallate, a water-insoluble phenolic compound with antioxidant, antibacterial, and water-repellent properties, onto wool. Grafting NDGA on wool with laccase produces multifunctional textiles with higher shrink resistance, tensile strength, and UV protection. Laccase may be utilized to create phenolic resins in situ, and enzymatic phosphorylation utilizing kinases can impart flame-retardant properties to natural fabrics [74, 99].

Similarly, alkaline pectinase, alpha-amylase, or laccase attached to textiles results in antimicrobial materials that keep full activity for at least ten wash cycles [35]. Two commercial enzymes, Termamyl® 2X (a-amylase) and Bioprep® 3000 L (alkaline pectinase), were immobilized on cotton fabric for antibacterial properties. Bioprep® 3000L enzyme exhibits higher antimicrobial activity against various bacteria than Termamyl® 2X, indicating that it has a lot of promise for replacing synthetic chemicals in antimicrobial textile manufacture [19].

In a novel technique, enzyme-functionalized cotton/polyethylene terephthalate (PET) blends are combined with eugenol-loaded human serum albumin (HSA)/silk fibroin (SF) nanocapsules to obtain antibacterial capabilities in smart antiseptic wound dressings [80]. The lysozyme was immobilized on the surface of wool fiber utilizing a Mannich reaction using THP as a cross-linker in recent research. When compared to untreated wool fabric, lysozyme-immobilized wool fabric demonstrated dramatically better antibacterial characteristics, with bacteria on wool–THP–lysozyme fabric decreasing by 95.38%. Wool–THP–lysozyme fabric still had effective bacteriostatic characteristics after five washing (73.31%) [101].

12 Enzyme Immobilized Textiles

Textiles as an excellent support material for biocatalyst immobilization in industrial applications. The various intrinsic properties of textiles allow them to design suitable for the specific application. Due to their availability in different forms, large surface area, high porosity, pore size, strength, simple preparation/functionalization procedures, and low-cost textiles have been a popular choice for a range of applications. The following benefits of using textile as a support matrix for immobilization of enzymes have been reported in previous literature [41, 59, 71, 92].

- Inexpensive availability
- Flexible construction (surface area, porosity)
- High efficiency in enzyme loading
- Robust system
- Process engineering
- Sustainability.

They also separate the immobilized catalysts from the reaction mixture perfectly and quickly, leaving no residue after usage. Furthermore, the textile surface's open, active, and porous character provides for low-pressure drop, allowing for fast reaction mixture dispersion, mass transfer, and excellent substrate turnover.

Since its inception in 1979 for biocatalyst immobilization on textiles, practically all types of textiles, including fibers, yarns, woven, non-woven, and knitted fabrics, have been employed for enzyme immobilization. Nonetheless, in addition to all basic forms of textiles, there is a large range of 3-D and intricate multilayered textile constructions for immobilization. Several enzymes have been included in a range of natural and synthetic textiles as support materials (including cotton, viscose, wool, silk, polyester, polyamide, and acrylic). For a strong enzyme/matrix interaction, textile surfaces were often functionalized with various cross-linking/binding polymers to provide functional groups such as amine/amide, aldehyde, hydroxyl, and carboxylic acid [41, 59].

An increasing interest in the usage of synthetic textiles has been noted among natural and synthetic ones. Their characteristics stay consistent over time, and their low cost makes them readily available and inexpensive.

Furthermore, their high chemical resistance combined with excellent processability gave researchers more freedom in developing a suitable surface for the immobilization of specific enzymes for a wide range of applications, from chemical synthesis to green and sustainable chemistry [41]. After immobilization on textiles, the exceptional features of various enzymes such as lipases, laccases, amylases, glucose oxidases, or peroxidases may be easily utilized for various applications in textile, food, beverage, detergent, medicines, and green energy [17, 59, 91, 92].

The immobilization of enzymes on textiles is governed by several interrelated factors that influence the performance of immobilized enzymes as well as the effectiveness of the immobilization process.

- Textile support matrix attributes (matrix form, total fiber surface area, shape, density, porosity, pore size distribution, and their operational stability)
- Binding interaction between textile and enzyme (nature of cross-linkers)
- Enzymatic activity stability and retention (functional groups on textile surface and the micro-environmental conditions)
- Enzyme orientation and multipoint attachments on the textile surface
- The extent of enzyme loading
- Mass transfer and diffusion limitations
- The lifetime of the immobilized enzyme
- Cost of both the enzyme and textile material.

The properties and features of the enzyme immobilized textiles are significantly influenced by the immobilization technique. Physical adsorption, covalent binding, cross-linking, and encapsulation are the most common methods of enzyme immobilization while novel techniques digital printing (inkjet and valve-jet) and electrostatic binding (layer-by-layer deposition, electrochemical doping/polymerization) also have been utilized in the design and development of enzyme immobilized textiles [59]. In a work this innovative approach offers the immobilization of lipase from thermolysis lanuginosus on polyethyleneimine treated cotton flannel cloth through layer-by-layer self-assembly technique and provides higher enzyme attachment to fabric [39]. The layer-by-layer assembly of controllable enzyme layers loaded with a high density of enzyme molecules through electrostatic deposition has been recently reported [104].

Inkjet printing is also accurate and fast enough to compete as an alternative method for enzyme immobilization, but vital considerations needed to be studied before the implementation of this approach. In recent work, using a digital inkjet printing system the oxidoreductase enzymes were immobilized on flexible textile fabric. Biswas studied the effectiveness of various factors in the immobilization of glucose oxidase and peroxidase on synthetic PET textile surfaces by this technique. In another study, the glucose-oxidase enzyme was directly printed on plasma-activated polyester fabric in a predefined pattern by valve-jet printing. They reported the enzyme printing concept as a resource-efficient enzyme immobilization strategy for sustainable applications, for example, controlled-release and biosensing [9].

12.1   Enzyme Immobilized Textiles in Wastewater Treatment

Immobilized enzymes have several benefits over native enzymes, particularly in terms of better stabilities (pH, heat, and storage), ease of separation, and reusability, as stated in previous sections. As summarized in Table 4, immobilized enzymes on textile are beneficial in various applications such as food preparation and preservation, medicines and medical devices, fine chemical synthesis, biodegradation, bioremediation, wastewater treatment applications, and so on [59, 71, 103]. Several papers have highlighted the promise of these enzyme immobilized textiles as a green and resource-efficient approach for removing poisonous colors and other impurities from water, based on a wide range of applications. The use of textile immobilized enzymes in textile wastewater treatment including the discoloration or removal of harmful dyes or colorants is the focus of this chapter.

According to [6], immobilized Horseradish peroxidase can remove 98% of azo dyes from wastewater in 45 min. They claim that the immobilized enzyme may be reused up to ten times while still keeping 69.6% of its initial activity. Commercial laccase was immobilized on coconut fibers, and a cheap and efficient biocatalyst for the repeated removal of numerous reactive dyes (individually and in combination) was produced [24]. When dyes are handled individually, they can be removed up to 90% of the time, but when colors are combined, they can be removed up to 63% of the time. Kahoush et al. [38] used a bio-Fenton and bio-electron Fenton system to study the elimination of reactive dyes utilizing immobilized glucose oxidase enzyme on carbon felt. Immobilization of enzymes, techniques/methods, performance factors, and overall utilization of immobilized enzymes in textile chemical processing have all been thoroughly examined in order to find gaps and prospective improvement opportunities for researchers in this field. Finally, the challenges and potential presented by this new technology are briefly examined.

13   Challenges in the Application of Immobilized Enzymes in the Textile Industry

Immobilization alters the characteristics of enzymes and is linked to insolubility and an increase in their size. The total or partial loss of catalytic activity during immobilization is one of the most serious issues related to enzyme immobilization. Various phenomena such as enzyme leaking during application, active site blockage owing to disoriented immobilization, and severe conditions during immobilization can cause immobilized enzymes to lose their activity. The significant constraint in textile chemical processing applications is due to the reduced diffusion of immobilized enzymes into the textile substrate compared to native enzymes. Because textiles are heterogeneous and macromolecular substrates, insoluble substrates have poor interactions with immobilized enzymes, resulting in low catalytic efficiency [46, 69, 92]. The performance of both the immobilization procedure and the catalytic application is

Table 4 Summary of enzyme immobilized on textiles with primary application areas

Enzyme	Immobilization method	Textile matrix	Stability/Performance	Application areas
Laccases	Covalent immobilization	Coconut fiber matrix	Improved thermal stability	Decolorization of dyes
		Poyacrylonitrile nanofibrous matrix	Improved storage stability with reusability	
		Electrospun chitosan/polyvinyl alcohol composite nanofibrous	Exhibit higher stability and reusability	Removal of phenol from wastewater
		Polyacrylic acid/SiO$_2$ nanofibrous matrix	Better storage stability with tolerance to pH and temperature	Triclosan removal from wastewater
		Polyacrylonitrile-biochar composite nanofibrous textiles	Improved storage, temperature and pH stability with reusability	Pharmaceutical compounds degradation
	Cross-linking immobilization	Polyacrylonitrile-biochar composite nanofibrous textiles	Improved storage, temperature, and pH stability	Degradation of chlortetracycline in continuous mode
		Polyamide 6/chitosan nanofibrous textile	Efficient for three treatment cycles	Removal of bisphenol and ethinylestradiol
	Encapsulation immobilization	Electrospun fibrous textile	Better storage and operational stability	Biodegradation of wastewater
	Physical adsorption	Coconut fiber matrix	Thermal and operational stabilities	Decolorization of dyes
Horseradish peroxidase	Covalent immobilization	Polyester, Polyamide, Viscose fabrics	Improved storage stability with retained activity up to 35 cycles	Biodegradation of wastewater

(continued)

Table 4 (continued)

Enzyme	Immobilization method	Textile matrix	Stability/Performance	Application areas
		Polymethyl methacrylate nanofibrous textiles	Improved storage stability and retained activity for reuse	Diagnostic and biosensors
		Polyvinyl alcohol/Polyacrylic acid/SiO_2 electrospinning nanofibrous textiles	Better storage stability with adequate reusability	Wastewater treatment
		Polyvinyl alcohol/Poly acrylamide nanofibrous textiles	Effective for 25 cycles of reuse	Conversion of phenol in wastewater treatment
		Polyester needle felt	Significant activity after 50 reuse cycles	Food processing (bio-bleaching of colored whey)
	Cross-linking immobilization	Polymethyl methacrylate Nanofibrous matrix	Retained 60% activity after 10 reuses	Biodegradation, bioremediation and Environmental application
		Polyvinyl alcohol/bovine serum albumin nanofibrous textiles	Retained 73% activity after 11 reuses	Food application, Bioremediation
	Encapsulation immobilization	Poly (d,l-lactide-co-glycolide)/PEO-PPO-PEO (f108) electrospun fibrous textiles	Better stability towards temperature and humic acids with storage stability	Removal of phenolic contaminants
	Physical adsorption	Acrylic textile	Improved stability toward pH, temperature, metal ions, and organic solvents	Removal of phenol from wastewater
		Wool textile	Improved thermal stability with higher resistance to metal ions, urea, detergent, proteolytic enzymes	Biodegradation, bioremediation, and environmental application

(continued)

Table 4 (continued)

Enzyme	Immobilization method	Textile matrix	Stability/Performance	Application areas
Catalase	Covalent immobilization	Cotton fabric	Reusability with excellent performance	Pharmaceuticals and fine chemical synthesis
	Covalent immobilization	Silver nanoparticle coated cotton fabric	Improved thermal stability	Multiple applications
Alcohol dehydrogenase	Covalent immobilization	Polyvinyl alcohol textiles	Better operational, pH, and thermal stability	Food preparation and preservation
Lipase	Covalent immobilization	Silk, cotton, polypropylene, polyethylene terephthalate, viscose textiles	Operational stability up to 35 cycles, viscose produce the most effective biocatalyst	Pharmaceuticals and fine chemical synthesis
	Cross-linking immobilization	Polyvinylalcohol/alginate, polyethylene oxide/alginate nanofibrous	Improved pH and thermal stability with retained activity after 14 reuses	Biodiesel production, transesterification, aminolysis, alcoholysis
Organophosphate hydrolase	Covalent immobilization	Polyester non-woven	Improve stability and performance	Environmental remediation of pesticides
Inulinase	Cross-linking immobilization	Polyester non-woven	Higher thermal stability with 38 cycles of reuses	Oxyfunctionalization of organic substrate
Glucose oxidase	Cross-linking immobilization/ Physical adsorption	Carbon felt	Better stability as well as good electrical and electrochemical characteristics	Biodegradation, diagnostic, and biosensors
	Physical adsorption	Polyester non-woven textiles	Improved thermal stability and retained 50% activity after 6 reuses	Oxyfunctionalization of organic substrate
α-Amylase	Physical adsorption	Acrylic textiles	Exhibit improved heavy metal, pH, and thermal stability	Food preparation and preservation

(continued)

Table 4 (continued)

Enzyme	Immobilization method	Textile matrix	Stability/Performance	Application areas
β-Galactosidase	Physical adsorption	Polyester non-woven	Retained potential for 15 reuse cycles	Food preparation and preservation
Trypsin	Cross-linking immobilization	Polyester fabric with carbodiimide, glutaraldehyde activation	Storage stability up to 20 days	Multifunction applications
		Polyester Non-woven with chitosan and glutaraldehyde activation	Improved activity with 20 days storage stability	Multifunction applications

influenced by mass transfer and diffusion limitations. Furthermore, different chemicals and auxiliaries are used in textiles, which frequently destabilize enzymes and therefore decrease their action. Nanoparticles and smart polymers as support have been offered as potential answers to these difficulties in textiles. Enzymes coupled to nano-size materials or smart polymers (soluble–insoluble) decreased mass transfer restrictions with the added benefit of in-depth understanding [46, 50].

There is no universal technique or support for immobilization due to vast differences in the characteristics of different enzymes and variable industry-specific needs for reaction processes. The ultimate support and immobilization technique are determined by the final practical environment. As a result, a considerable study should be concentrated on enzyme immobilization techniques connected to textile chemical processing to broaden their horizons for textile applications.

14 Conclusion

Enzymes are widely sought after for catalyzing different chemical reactions in textiles and ensuring a greener manufacturing operation. The prospects and advancement of sustainability and green technology in various domains of application are expanding tremendously in this era of sustainability and green technology. Enzymes are applied in textile processing at mild conditions, resulting in lower greenhouse gas emissions and less reliance on non-renewable resources. Enzymatic biotechnology has promise for improving sustainability principles in the textile manufacturing supply chain in today's competitive and demanding business environment.

However, the limitations in biocatalyst stability and adequate reusability largely hampered the scale-up of enzymatic processes. The immobilization process, as mentioned in this study, involves recyclability and reusability, which provides high stability and appropriate activity to enzymes, allowing for improved control of the process and a reduction in chemical, energy, and time consumption, lowering the costs. Enzyme immobilization has been proposed and researched as a beneficial approach for broadening the scope of enzyme catalysis and improving process efficiency to achieve this goal.

The structure–function relationship of the enzyme and the platform used in immobilization, as well as the optimized product yield at low implementation costs, and an integrated combinatorial strategy that combines experimental and computational approaches, will all play a role in future implementation prospects. Because of the technological constraints of adopting immobilized enzymes, producing immobilized biocatalysts for commercial use will necessitate a multidisciplinary approach. The overlap of technical expertise in enzyme immobilization, protein engineering, process engineering, and life cycle analysis is critical for the next generation of immobilized enzymes and their efficient commercial-scale deployment.

References

1. Al-Bar OAM, El-Shishtawy RM, Mohamed SA (2021) Immobilization of camel liver catalase on nanosilver-coated cotton fabric. Catalysts 11(8):900. https://doi.org/10.3390/catal11 1080900
2. Ali S, Zafar W, Shafiq S, Manzoor M (2017) Enzymes immobilization: an overview of techniques, support materials and its applications. Int J Sci Technol Res 6(7):64–72
3. Amari A, Alzahrani FM, Alsaiari NS, Katubi KM, Rebah FB, Tahoon MA (2021) Magnetic metal organic framework immobilized laccase for wastewater decolorization. Processes 9(5):774. https://doi.org/10.3390/pr9050774
4. Amorium AM, Gasques MDG, Andreaus J, Scharf M (2002) The application of catalase for the elimination of hydrogen peroxide residues after bleaching of cotton fabrics. Annals of the Brazilian Academy of Sciences. 74(3):433–436
5. Arputharaj A, Raja ASM, Saxena S (2016) Developments in sustainable chemical processing of textiles. In: Muthu SS, Gardetti MA (eds) Green fashion. Environmental footprints and eco-design of products and processes. Springer, pp 217–252. https://doi.org/10.1007/978-981-10-0111-6_9
6. Arslan M (2011) Immobilization horseradish peroxidase on amine-functionalized glycidyl methacrylate-g-poly (ethylene terephthalate) fibers for use in azo dye decolorization. Polym Bull 66(7):865–879
7. Aslam S, Asgher M, Khan NA, Bilal M (2021) immobilization of pleurotus nebrodensis WC 850 laccase on glutaraldehyde cross-linked chitosan beads for enhanced biocatalytic degradation of textile dyes. J Water Process Eng 40:101971. https://doi.org/10.1016/j.jwpe.2021.101971
8. Besegatto SV, Costa FN, Damas MSP, Colombi BL, De Rossi AC, Aguiar CRL, Immich APS (2018) Enzyme Treatment at different stages of textile processing: A Review. Ind Biotechnol 14(6):298–307. https://doi.org/10.1089/ind.2018.0018
9. Biswas T, Yu J, Nierstrasz V (2021) Effective pretreatment routes of polyethylene terephthalate fabric for digital inkjet printing of enzyme. Adv Mater Interf 8(6). https://doi.org/10.1002/admi.202001882
10. Blackburn RS (2009) Sustainable textiles Life cycle and environmental impact. Woodhead Publishing Limited, Cambridge, UK, The Textile Institute
11. Brena B, Gonzalez-Pambo P, Batista-Viera F (2013) Immobilization of enzymes: a literature survey. In: Guisan JM (ed) Immobilization of enzymes and cells, 2nd edn. Humana Press Inc, Totowa, NJ, pp 15–30
12. Cao L (2005) Carrier-bound immobilized enzymes: Principles, applications and design. Wiley-Vch Verlag GmbH & Co, KGaA, Weinheim
13. Capecchi E, Piccinino D, Bizzarri BM, Avitabile D, Pelosi C, Colantonio C, Calabrò G, Saladin R (2019) Enzymes-lignin nanocapsules are sustainable catalysts and vehicles for the preparation of unique polyvalent bio-inks. Biomacromolecules 20. https://doi.org/10.1021/acs.biomac.9b00198
14. Chang Y, Yang D, Li R, Wang T, Zhu Y (2021) Textile dye biodecolorization by manganese peroxidase: A review. Molecules 26(15):4403. https://doi.org/10.3390/molecules26154403
15. Chapman J, E.Ismail A, Dinu CZ (2018) Industrial applications of enzymes: recent advances, techniques, and outlooks. Catalysts. 8:238. https://doi.org/10.3390/catal8060238
16. Chen JY, Sun L, Edwards VJ (2014) Regenerated cellulose fiber and film immobilized with lysozyme. Bioceram Dev Appl 4(1). https://doi.org/10.4172/2090-5025.1000078
17. Chibata I (1996) Industrial applications of immobilized biocatalysts and biomaterials. In: advances in molecular and cell biology, Elsevier Publishing, pp 151–160
18. Choudhury AK (2014) Sustainable textile wet processing: Applications of enzymes. In: Muthu S (ed) Roadmap to sustainable textiles and clothing: textile science and clothing technology. Springer, Singapore. https://doi.org/10.1007/978-981-287-065-0_7

19. Coradi M, Zanetti M, Valério A, de Oliveira D, da Silva A, de Souza SMAGU, de Souza AAU (2018) Production of antimicrobial textiles by cotton fabric functionalization and pectinolytic enzyme immobilization. Mater Chem Phys 208:28–34. https://doi.org/10.1016/j.mat chemphys.2018.01.019

20. Costa SA, Tzanov T, Paar A, Gudelj M, Gubitz GM, Cavaco-Paulo A (2001) Immobilization of catalases from Bacillus SF on alumina for the treatment of textile bleaching effluents. Enzyme Microbial Technology. 28:815–819

21. Costa SA, Tzanov T, Paar A, Carneiro F, Gubitz GM, Cavaco- Paulo A (2002) Recycling of textile bleaching effluents for dyeing using immobilized catalase. Biotech Lett 24:173–176

22. Costa C, Azoia N G, Silva C, Marques EF (2020) Textile industry in a changing world: challenges of sustainable development. U Porto J Eng 6(2):86–97. https://doi.org/10.24840/ 2183-6493_006.002_0008

23. Courth K, Binsch M, Ali W, Ingenbosch K, Zorn H, Hoffmann-Jacobsen K, Gutmann JS, Opwis K (2021) Immobilization of peroxidase on textile carrier materials and their application in the bleaching of colored whey. J Dairy Sci 104(2):1548–1559. https://doi.org/10.3168/jds. 2019-17110

24. Cristovao RO, Silverio SC, Tavares APM, Brígida AIS, Loureiro JM, Boaventura RAR, Macedo EA, Coelho MAZ (2012) Green coconut fiber: a novel carrier for the immobilization of commercial laccase by covalent attachment for textile dyes decolourization. World J Microbiol Biotechnol 28(9):2827–2838

25. Datta S, Christena LR, Rajaram YRS (2013) Enzyme Immobilization: an overview on Techniques and Support Materials. Biotech 3(1):1–9

26. Dettore C (2011) Comparative life-cycle assessment (LCA) of textile bleaching systems: gentle power bleach vs. conventional bleaching System. In: Scientific poster, AATCC international conference, pp 22–24 March

27. Dincer A, Telefoncu A (2006) Improving the stability of cellulose by immobilization on modified polyvinyl alcohol-coated chitosan beads. J Molecular Catalysis B: Enzymatic 45:10–14

28. Dourado F, Bastos M, Mota M, Gama FM (2002) Studies on the properties of Celluclast/Eudragit L-100 conjugate. J Biotechnol 99:121–131

29. Eid MB, Ibrahim NA (2021) Recent developments in sustainable finishing of cellulosic textiles employing biotechnology. J Clean Prod 284. https://doi.org/10.1016/j.jclepro.2020.124701

30. Galante YM, Cristina F (2003) Enzyme applications in detergency and manufacturing industries. Curr Org Chem 7(13):1399–1422

31. Gokalp N, Ulker C, Guvenilir YA (2016) Enzymatic ring opening polymerization of caprolactone by using a novel immobilized biocatalyst. Adv Mater Lett 7(2):144–149. https://doi. org/10.5185/amlett.2016.6059

32. Gulzar T, Farooq T, Kiran S, Ahmad I, Hameed A (2019) Green chemistry in the wet processing of textiles. In: Shahid-ul-Islam, Butola BS (eds) The impact and prospects of green chemistry for textile technology, Woodhead Publishing, ISBN 9780081024911, pp 1–20

33. https://sustainablecampus.fsu.edu/blog/clothed-conservation-fashion-water assessed on 15th Oct, 2021

34. https://economicsofwater.weebly.com/water-usage-and-the-textile-industry.html assessed on 15th Oct, 2021

35. Ibrahim NA, Gouda M, El-Shafei AM, Abdel-Fatah OM (2007) Antimicrobial activity of cotton fabrics containing immobilized enzymes. J Appl Polym Sci 104:1754–1761. https:// doi.org/10.1002/ap.25821

36. Jegannathan KR, Nielsen PH (2013) Environmental assessment of enzyme use in industrial production- a literature review. J Clean Prod 42:228–240

37. Kabir SMF, Chakraborty S, Hoque SMA, Mathur K (2019) Sustainability assessment of cotton-based textile wet processing. J. of Cleaner Technology. 1(1):232–246. https://doi.org/ 10.3390/cleantechnol1010016

38. Kahoush M, Behary N, Guan J, Cayla A, Mutel B, Nierstrasz V (2021) Genipin-mediated immobilization of glucose oxidase enzyme on carbon felt for use as heterogeneous catalyst

in sustainable wastewater treatment. J Environ Chem Eng 9(4). https://doi.org/10.1016/j.jece. 2021.105633

39. Karimpil JJ, Melo JS, D'Souza SF (2012) Immobilization of lipase on cotton cloth using the layer-by-layer self-assembly technique. Int J Biol Macromol 50(1):300–302
40. Khan AA, Alzohairy MA (2010) Recent advances and applications of immobilized enzyme technologies: A review. Research Journal of Biological Sciences 5(8):565–575
41. Kiehl K, Straube T, Opwis K, Gutmann JS (2015) Strategies for permanent immobilization of enzymes on textile carriers. Eng Life Sci 15:622–626. https://doi.org/10.1002/elsc20140 0148
42. Kołodziejczak-Radzimska A (2021) Nghiem LD & Jesionowski T (2021) Functionalized Materials as a Versatile Platform for Enzyme Immobilization in Wastewater Treatment. Curr Pollution Rep 7:263–276. https://doi.org/10.1007/s40726-021-00193-5
43. Krikstolaitytea V, Kuliesiusa J, Ramanavicieneb A, Mikoliunaitea L, Minkstimieneb AK, Oztekina Y, Ramanavicius A (2014) Enzymatic polymerization of polythiophene by immobilized glucose oxidase. Polymer 55(7):1613–1620. https://doi.org/10.1016/j.polymer.2014. 02.003
44. Krishnamoorthi S, Banerjee A, Roychoudhury A (2015) Immobilized Enzyme Technology: Potentiality and Prospects. J Enzymol Metabol. 1(1):104
45. Kumar VS, Meenakshisundaram S, Selvakumar N (2008) Conservation of cellulase enzyme in biopolishing application of cotton fabrics. The Journal of Textile Institute. 99(4):339–346
46. Kundu D, Thakur MS, Patra S (2021) Textile fabric processing and their sustainable effluent treatment using enzymes—insights and challenges. In: Tripathi A, Melo JS (eds) Immobilization strategies: biomedical, bioengineering and environmental applications, Springer, pp 645–666. https://doi.org/10.1007/978-981-15-7998-1
47. Lellis B, Fávaro-Polonio CZ, Pamphile JA, Polonio JC (2019) Effects of textile dyes on health and the environment and bioremediation potential of living organisms. Biotechnology Research and Innovation 3:275–290. https://doi.org/10.1016/j.biori.2019.09.001
48. Liese A, Hilterhaus L (2013) Evaluation of immobilized enzymes for industrial applications. Chem Soc Rev 42(15):6236–6249. https://doi.org/10.1039/C3CS35511J
49. Lopes LA, Dias LP, Silva da Costa HP, da Silva X, Neto J, Morais EG, Abreu de Oliveira JT, Vasconcelos IM, de Oliveira D, de Sousa B (2021) Immobilization of a peroxidase from Moringa oleifera Lam. roots (MoPOX) on chitosan beads enhanced the decolorization of textile dyes. Process Biochem 110:129–141. https://doi.org/10.1016/j.procbio.2021.07.022
50. Madhu A, Chakraborty JN (2017) Developments in application of enzymes for textile processing: Journal of Cleaner Production. 145:114–133
51. Madhu A, Chakraborty JN (2018) Recover and reuse of α- amylase enzyme for cotton fabric desizing using immobilization. Res J Textile Appar 22 (3):271–290. https://doi.org/10.1108/ RJTA-12-2017-0052
52. Madhu A, Chakraborty JN (2019) Bio-Bleaching of Cotton with H_2O_2 generated from native and immobilized glucose oxidase. AATCC J Res 6(2):7–17. https://doi.org/10.14504/ajr.6.2.2
53. Martorana A, Bernini C, Valensin D, Sinicropi A, Pogni R, Basosi R, Baratto MC (2011) Insights into the homocoupling reaction of 4-methylamino benzoic acid mediated by Trametes versicolor laccase. Mol BioSyst 7:2967–2969
54. Mateo C, Palomo JM, Fernandez-Lorente G,Guisan JM, Fernandez-Lafuente R (2007) Improvement of enzyme activity, stability and selectivity via immobilization techniques. Enzym Microb Technol 40: 1451–1463. https://doi.org/10.1016/j.enzmictec.2007.01.018.
55. Mendoza-Avila J, Chauhan K, Vazquez-Duhalt R (2020) Enzymatic synthesis of indigo-derivative industrial dyes. Dye Pigment 178:108384 https://doi.org/10.1016/j.dyepig.2020. 108384
56. Miletic N, Nastasovic A, Loos K (20s12) Immobilization of biocatalysts for enzymatic polymerization: possibilities, advantages, applications. Biores Tech 115:126–135. https://doi.org/ 10.1016/j.biortech.2011.11.054

57. Mohamad NR, Marzuki NHC, Buang NA, Huyop F, Wahab RA (2015) An overview of technologies for immobilization of enzymes and surface analysis techniques for immobilized enzymes. Biotech & Biotechno Equip 29(2):205–220. https://doi.org/10.1080/131 02818.2015.1008192

58. Mojsov KD (2014) Trends in Bio-processing of Textiles: A review. Advanced Technologies 3(2):135–138

59. Morshed MN, Behary N, Bouazizi N, Guan J, Nierstrasz VA (2021) An overview on biocatalysts immobilization on textiles: preparation, progress and application in wastewater treatment. Chemosphere 279:130481. https://doi.org/10.1016/j.chemosphere.2021.130481

60. Morsy SAGZ, Tajudin AA, Ali MSMA, Sharif FM (2020) Current development in decolorization of synthetic dyes by immobilized laccases. Front Microbiol 11:572309. https://doi.org/10.3389/fmicb.2020.572309

61. Muthu SS (2018a) Sustainable innovations in textile chemical processes. Textiles and clothing sustainability, textile science and clothing technology, Springer, Singapore

62. Muthu SS (2018b) Sustainable innovations in recycled textiles. Textiles and clothing sustainability, textile science and clothing technology, Springer, Singapore

63. Muthu SS (2019) Water in textiles and fashion: consumption, footpring and life cycle assesment. Woodhead Publishing, UK

64. Muthu SS (2020) Assessing the environmental impact of textiles and the clothing supply chain, 2nd edn. Woodhead Publishing, United Kingdom, The Textile Institute Book Series. https://doi.org/10.1016/B978-0-12-819783-7.00001-6

65. Nadaroglu H. (2021) Immobilization and application of industrial enzymes on plant-based new generation polymers. In: Malik S (ed) Exploring plant cells for the production of compounds of interest. Springer, Cham. https://doi.org/10.1007/978-3-030-58271-5_9.

66. Nayak R, Singh A, Panwar T, Padhye R (2019) A review of recent trends in sustainable fashion and textile production. Current Trends Fash Technol Textile Eng 4(5) https://doi.org/10.19080/CTFTTE.2019.04.555648

67. Nielsen PH, Skagerlind P (2007) Cost-neutral replacement of surfactants with enzymes. Household and Personal Care Today 4:3–7

68. Nielsen PH, Kuilderd H, Zhou W, Lu X (2009) Enzyme biotechnology for sustainable textiles. In: Blackburn RS (ed) Sustainable textile: life cycle and environment impact. Woodhead Publishing Limited, CRC Press, pp 113–138

69. Nierstrasz VA, Cavaco-Paulo A (eds) (2010) Advances in textile biotechnology. Woodhead Publishing Ltd., Cambridge

70. Norouzian D (2003) Enzyme immobilization: the state of art in biotechnology. Iranian J of Biotech 1(4):197–206

71. Opwis K, Straube T, Kiehl K, Gutmann JS (2014) Various strategies for the immobilization of biocatalysts on textile carrier materials. Chem Eng Trans 38:223–228. https://doi.org/10.3303/CET1438038

72. Osiadacz J, Al-Adhami AJH, Bajraszewska D, Fischer P, Peczyn̄ska-Czoch W. (1999). On the use of trametes versicolor laccase for the conversion of 4-methyl-3 hydroxyanthranilic acid to actinocin chromophore. J Biotechnol 72:141–149

73. Parisi ML, Fatarella E, Spinelli D, Pogni R, Basosi R (2015) Environmental impact assessment of an eco-efficient production for coloured textiles. J Clean Prod 108:514–524. https://doi.org/10.1016/j.jclepro.2015.06.032

74. Paul R, Genesca E (2013) The use of enzymatic techniques in the finishing of technical textiles. In: Gulrajni ML (ed) Advances in dyeing and finishing of technical textiles. Woodhead Publishing Ltd., Cambridge, pp 177–198

75. Pazarlioglu NK, Sariisik M, Telefoncu A (2005) Treating denim fabrics with immobilized commercial Cellulases. Process Biochem 40:767–771

76. Prakash O, Jaiswal N (2011) Immobilization of a thermostable amylase on agarose and agar matrices and its application in starch stain removal. World Appl Sci J 13(3):572–577

77. Polak J, Jarosz-Wilkolazka A (2012) Fungal laccases as green catalysts for dye synthesis. Process Biochem 47:1295–1307. https://doi.org/10.1016/j.procbio.2012.05.006

78. Polak J, Jarosz-Wilkołazka A (2010) Whole-cell fungal transformation of precursors into dyes. Microb Cell Fact 9:51–62. http://www.microbialcellfactories.com/content/9/1/51

79. Popović N, Stanišić M, Đurđić KI, Prodanović O, Polović N, Prodanović R (2021) Dopamine modified pectin for a Streptomyces cyaneus laccase induced microbeads formation, immobilization, and textile dyes decolorization. Environ Technol Innov 22:101399. https://doi.org/10.1016/j.eti.2021.101399

80. Quartinello F, Tallian C, Auer J, Scho H, Vielnascher R, Weinberger S, Wieland K, Weihs AM, Herrero Rollett A, Lendl B, Teuschl AH, Pellis A, Guebitz GM (2019) Smart textiles in wound care: functionalization of cotton/PET blends with antimicrobial nanocapsules. J Mater Chem B 7:6592–6603. https://doi.org/10.1039/c9tb01474h

81. Sahinbaskan BY, Kahraman MV (2011) Desizing of untreated cotton fabric with the conventional and ultrasonic bath procedures by immobilized and native α-amylase. Starch 63(3):154–159. https://doi.org/10.1002/star.201000109

82. Saxena S, Raja ASM, Arputharaj A (2017) Challenges in sustainable wet processing of textiles In: Muthu S (ed) Textiles and clothing sustainability. Textile science and clothing technology. Springer, Singapore. https://doi.org/10.1007/978-981-10-2185-5_2

83. Schroeder M, Schweitzer M, Lenting HBM, Guebitz GM (2004) Chemical modification of protease for wool cuticle scale removal. Biocatal Biotransform 22(5/6):299–305

84. Shah T & Halacheva S (2015) Drug-releasing textiles. In: Langenhove L (ed) Advances in smart medical textiles: treatments and health monitoring. Woodhead Publishing Ltd., Cambridge, pp 119–154

85. Sheikh J, Bramhecha I (2019) Enzymes for green chemical processing of cotton In: Shahid-ul-Islam, Butola BS (ed) The impact and prospects of green chemistry for textile technology. The textile institute book series, Woodhead Publishing, pp 135–160. https://doi.org/10.1016/B978-0-08-102491-1.00006-X

86. Shekh MdMK, Koh J (2021) Sustainable textile processing by enzyme applications. Intechopen Publication. https://doi.org/10.5772/intechopen.97198

87. Sheldon RA (2007) Enzyme immobilization: The quest for optimum performance'. Adv Synth Catal 349:1289–1307

88. Shen J, Rushforth M, Cavaco-Paulo A, Guebitz G, Lenting H (2007) Development and industrialization of enzymatic shrink-resist process based on modified proteases for wool machine wash-ability. Enzyme Microb Technol 34:1–6

89. Shen J (2010) Enzymatic treatment of wool and silk fibers. In: Nierstrasz VA, Cavaco-Paulo A (ed). Advances in textile biotechnology. Woodhead Publishing Ltd., Cambridge, pp 171–192

90. Silva CJSM, Zhang Q, Shen J, Cavaco-Paulo A (2006) Immobilization of proteases with a water-soluble–insoluble reversible polymer for treatment of wool. Enzyme Microb Technol 39:634–640

91. Singh RS, Singhania RR, Pandey A, Larroche C (eds) (2019) Advances in enzyme technology. Elsevier Publishing

92. Soares JC, Moreira PR, Queiroga AC, Morgado JE, Malcata FX, Pintado ME (2011) Application of immobilized enzymes technologies for the textile industry: A review. Biocatal Biotransform 29(6):223–237

93. Sousa AC, Oliveira MC, Martins LO, Robalo MP (2018) A sustainable synthesis of asymmetric phenazines and phenoxazinones mediated by CotA-Laccase. Adv Synth Catal 360(575):583. https://doi.org/10.1002/adsc.201701228

94. Talekar S, Joshi A, Joshi G, Kamat P, Haripurkar R, Kambale S (2013) Parameters in preparation and characterization of cross-linked enzyme aggregates (CLEAs). RSC Advance 3:12485–12511

95. Tripathi A, Melo JS (eds) (2021) Immobilization strategies: biomedical, bioengineering and environmental applications. Springer, pp 645–666. https://doi.org/10.1007/978-981-15-7998-1

96. Tzanov T, Calafell M, Guebitz G M, Cavaco-Paulo A (2001) Bio-preparation of cotton fabrics. Enzym Microbiol Technol 29:357–362.

97. Tzanov T, Costa SA, Gubitz GM, Cavaco-Paulo A (2002) Hydrogen peroxide generation with immobilized glucose oxidase for textile bleaching. J Biotechnol 93:87–94

98. Vasconcelos A, Silva CJSM, Schroeder M, Guebitz GM, Cavaco-Paulo A (2006) Detergent formulations for wool domestic washings containing immobilized enzymes'. Biotechnology Letter 28:725–731

99. Wehrschutz-Sigl E, Hasmann A, Guebitz GM (2010) Smart textiles and biomaterials containing enzymes or enzymes substrates In: Nierstrasz VA, Cavaco-Paulo A (eds) Advances in textile biotechnology. Woodhead publishing Ltd., Cambridge, pp 171–192

100. Wlizłoa K, Polaka J, Jarosz-Wilkołazkaa A, Pognib R, Petricci E (2020) Novel textile dye obtained through transformation of 2-amino-3-methoxybenzoic acid by free and immobilised laccase from a Pleurotus ostreatus strain. Enzym Microb Technol 132:109398 https://doi.org/10.1016/j.enzmictec.2019.109398

101. Yang W, Zhang N, Wang Q, Wang P, Yu Y (2020) Development of an eco-friendly antibacterial textile: lysozyme immobilization on wool fabric. Bioprocess Biosyst Eng. https://doi.org/10.1007/s00449-020-02356-y

102. Yu Y, Yuan J, Wang Q, Fan X, Ni X, Wang P, Cui L (2013) Cellulase immobilization onto the reversibly soluble methacrylate copolymer for denim washing. Carbohyd Polym 95(2):675–680

103. Zdarta J, Jankowska K, Bachosz K, Degórska O, Kaźmierczak K, Nguyen LN, Nghiem LD, Jesionowski T (2021) Enhanced Wastewater Treatment by Immobilized Enzymes. Curr Pollution Rep. 7:167–179. https://doi.org/10.1007/s40726-021-00183-7

104. Zhang J, Huang X, Zhang L, Si Y, Guo S, Su H, Liu J (2020) Layer-by-layer assembly for immobilizing enzymes in enzymatic biofuel cells. Sustainable Energy Fuels 4(1):68–79s

Dabu, The Sustainable Resist Printed Fabric of Rajasthan

Janmay Singh Hada and Chet Ram Meena

Abstract Sustainable design solutions require inventing solutions equally favourable to humanity and its ecology. The eco-design, choosing raw materials with regional character employing low carbon footprint is a critical factor of these mud resist prints. *'Dabu or Daboo'* is block printed Mud resisted traditional textiles in dark earthy tones of stunning designs and conventional patterns on natural fabrics in which carved wooden way on wooden/metal blocks are stamped using Mud resist print techniques. This chapter attempts to record the marvellous resist printed (natural colours) textile fabric of *Dabu or Daboo*, which has got the verge of extinction in its pure form. The reasons include faster printing methods, usage of chemical dyes and more painstaking method. The objective is documentation of the craft in terms of its sustainable measures and modifications that have come about in its technique, types of colours, motif and sprouts beautiful patterns which are handed down intact, over generations. The craft was collected purposively selected samples traditionally practising in the craft. It is different from Ajarakh prints in terms of lime replaced by black clay (Mud). Dabu is a precious wooden block of printed fabric that ought to be protected. The conservation needed precise documentation of the craftwork, its chronology, procedure, motifs, vibrant colours and by-products.

Keywords Dabu · Sustainability · Indigo dye · Resist printing · Mordant

1 Introduction

In India, textiles and costumes replicate the distinctive and significance traditions of regions and their ethnography. India has a centuries-old custom of using natural dyes for colouring textile materials, households foods items and skin care products,

J. S. Hada (✉) · C. R. Meena
Department of Textile Design, National Institute of Fashion Technology, Jodhpur, Rajasthan 342037, India
e-mail: janmay.hada@nift.ac.in

C. R. Meena
e-mail: chetram.meena@nift.ac.in

which eventually stepped into folk art traditions. Natural dyeing comprises those colourants obtained from natural sources like insects, animal and vegetable matter without chemical processing [25]. These regional traditional textiles are well-looked as an inherited gift by artisans but modify to different shapes and products by the rest of the people to meet their objective for new trendy sustainable fabric. Sustainability is an integrated sign of eco, green and ethical drive movement in society. It incites consciousness concerning the fashion impact on the environment augmented resources, attributed to the quality and value of textile clothing [13]. Sustainability is an integral part of the Dabu block printing process in which beautiful designs and patterns are stamped on fabric inspired by local flora and fauna. This fabric is aboriginal and has come to marks characterise the area. The mud resists dabu textiles prints by wooden blocks have the uniqueness of being subtle but remain so remarkable that they are part of a sustainable fashion chain. In the present sustainability scenario, eco-friendly and less carbon emission are buzz words in fashion clothing [17].

Textile printing is the innovative, adaptable and important method for colouring textile fabrics for value addition [3]. It is a process of bringing together a design idea, one or more colourants, with a substrate (usually a cloth), using a method for applying the colourants with some precision in any pattern or motifs [2].

When we look history, the cotton fragments unearthed from Egypt (Quesir-al-Qadim and Fostat, Al Fusta) seem to be earlies printed fabric from India [20]. It is localised dyeing to make exquisite configurations and patterns of the fabrics, i.e. restricted to certain parts of cloth that constitute design to create value and aesthetic effect on the materials from ancient time[4]. Earliest examples of applying colour in print date back to 3000 BC. Archaeological proof from the Indus valley civilisation (Mohenjo-Daro) ascertains technology of mordant dyeing had been known in the Asian subcontinent from at least the 2nd BC. The evolution of blocks and stencils offered consistency to this early form of decoration. Colouring materials are applied to the remaining raised area, thus providing design when stamped or pressed onto the fabric [1].

The term printing signifies the production by various means of coloured patterns or designs on textile material rather than woven, embroidered or painted designs. It is probably the cheapest method of ornamenting textile materials and is very popular because of its beautiful effect (Table 1).

The fabric is printed with one or more vibrant colours in particular sharply defined patterns instead the whole cloth is uniformly covered with one colour by exhaust process [5] by direct, discharge and resist methods. Nowadays, textile printing uses specific methods, techniques and machines (Screen printing, digital ink-jet printing, or usage of thermal transfer processes or transfer printing). Time, productivity, flexibility, creativity are unique features of contemporary methods in textile printing. [6] but traditional blocks prints possess artistic and decorative with purity and richness of colour is still very demandable and used by many consumers and fashion brands. Traditional craft reflects the cultural, socio-economic, historical behaviour of the society. So it has to be conserved with two propositions: The preservation from misuse and commercialisation and modification of craft by the contemporary environment. Dabu is the treasured block-printed textiles that need to be preserved. So

Table 1 Some Common materials used in dabu printing paste

S.No	Ingredients of printing paste	Scientific name
1	Harda: Myrobalan	Terminalia chebula
2	lokhand	Scrap Iron
3	Guar gum	Cyamopsis tetragonoloba
4	Dhaori Ka phool	Galls of tamarisk
5	Kuwadia na beej	Casatoria seeds
6	Sajikhar	Salt
7	Majith	Madder
8	Fitakadi	Alum
9	Haldi or turmeric	Curcuma domestica
10	Dadam or Anar	Pomegranate
11	Gud	Jaggery
12	Natural Indigo	Indigofera tinctoria
13	Ruharb (Brown Colour)	Rheum rhabarbarum

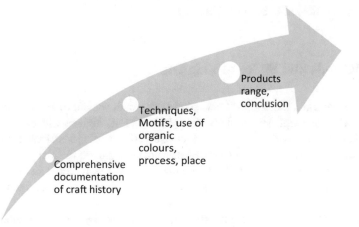

Products range, conclusion

Techniques, Motifs, use of organic colours, process, place

Comprehensive documentation of craft history

Fig. 1 Research roadmap for *Dabu,* the sustainable resist printed fabric

drawing out the entire process would be enormously useful and, more significantly, provide a path for the preservation for complete illustration and writeup about motifs, tool, techniques, colours and history (Fig. 1).

2 Methodology

The chapter aims at the documentation of *Dabu*; a Mud block printed textile of Rajasthan. To meet this objective, descriptive research was designed, for which an

Table 2 Composition of one kilogram Indigo vat used for making dabu hand block print paste

S.No	One kilogram indigo vat contains (1000 gm)	Weight (gm)
1	Indigo powder	1000
2	Caustic soda	250–300
3	Sodium hydro sulphite	1000–1500

Earlier vat made in clay pot or vessel of nearby pond or river (tanka) of 300–500-L capacity with lime, fenugreek, jiggery and millet flour.

open and close-ended questionnaire was structured and a snowball sampling method was followed. These queries dealt with the chronology of the craft, demographic details, process (both traditional and contemporary), motif, colour, products created. For the gathering of accurate data, interviews were scheduled with artisans, recording the discussion with a photograph.

The purposive method of sampling was used in the research. The two families were examined, who are doing this craftwork in Pipad and Barmer city in customary form. So two families were selected, first from the Pipad village of Jodhpur District and second units in the Barmer (Table 2).

3   Methods and Styles of Printing

Each state in India has several handloom crafts with special distinctive features. In the global trade, emphasis should be on colour, styling, high-quality design, function and user performance to make the product competitive. Methods of Printing mean the means of tools producing the printed effect depending upon the means employed. In Printing, different methods and tools are used, like wooden blocks in block printing, engraved rollers (emboss or roller prints), screens. Direct, resist and discharge are basic styles of printing for colour on fabrics [12].

The printing method is block, stencil, and screen (by hand or manual screen, automatic flatbed screen, rotary screen) roller printing and Heat transfer printing (sublistatic prints). In transfer, printing colour is transferred from the surface of the paper to the fabric. Under this high temperature, dye is submitted and is transferred onto fabric [11]. The above methods define moving the pattern to fabric with inherent advantages and disadvantages [4].

The thickener is added to the solution to increase the viscosity of the printing paste, which stops it from spreading (by capillary action beyond the limits of pattern or design). Organic colours and auxiliaries are added in printing paste to improve the fixation in fabric. After printing, the fabric is passed through a subsequent steaming process and after treatments before they are thoroughly developed and adjudicated fast to light and washing [7]. The same principle of specific dye classes having selecting fibre affinities and general fastness characteristics apply equally to Printing and dyeing [1]. The world 'natural dye' covers all the dyes obtained from natural sources like roots, stem, bark, leaves, fruits and seeds that may contain colouring matter extracted from different sources and methods [23]. The extraction efficiency of colourant depends on sources (Nature like mineral, flora and fauna) depending on the type of solvents (organic), acid, alkali, pH and extraction conditions like temperature, time, material to liquor ratio (M.L.R) and the size of particles [24].

The most common style of printing in block prints is the direct method. The white or coloured fabric (overprint) is used in the direct process. The printing paste is applied directly on the fabric surface without alterations. The discharge style depends on dyeing the fabric first and then printing with a chemical that will destroy the colour in the designed areas. Sometimes the base colour is removed, and another colour is printed, but white space is usually desirable to brighten the overall design. The coloured discharge is used by utilising dyestuff which is not reduced by removing chemicals. It is followed by a steaming process. In coloured discharged numerous effects can be accomplished on the materials.

In the Dabu style of printing (as its name implies), the pattern area is resisted by mud and then dye may be applied by exhaust process. The fabric is dyed on all surfaces excluding the parts obscured by resist paste. Resist styles diverged into different resist chemicals (those that employ agents such as glyoxal-bisulfite adducts or stannous chloride) and physical (those that use wax to block the fibre from being dyed) methods [2]. The natural colouring materials generate new scopes for researchers to replace synthetic dyes [10].

4 Dabu Hand Block Printing

Dabu means *'Dabana'*, which means press by Mud resist block prints. It mentions the techniques by which engraved blocks (wooden, metal-like t japs) covered with dye are repeatedly pressed along the fabric to create patterns. It is the ancientest and most uncomplicated method of printing. Block prints are believed to have originated in Asia (China, 2000 years old). In the Diamond sutra (868BC) it is mentioned, which is currently in the British Museum [8]. In Block prints, design is manipulated, making colourways put into repeat, and colourways create and separate colour from time to time [9]. The popular methods are Kahma, Kantedar, Lal Titri and Dholakia.

Earlier, the block makers were either from their community or the Suthar community. In Barmer, a local carpenter from the Suthar community is available, whereas in Pipad, mostly block made from Jaipur by craftspersons belong to Farukhabad, Uttar Pradesh. The skill and experience, as the craftsmen swiftly pattern the clothes in exquisite designs. The delicate line designs are to be carved in a wooden block using Cu (copper) stripes with metal pins. The design is a transfer by stamping the block using hand pressure (wooden mallet) on the cloth. They lifted and stamped on the cloth continually. The size and weight of blocks are light; so, it is easily used by artisans. Each new area of the fabric has to be printed by a lay-out of design, and successive impressions have to be adjusted accurately to each other. Each colour requires a distinct block. In large designs, each colour may require more than one block. In 'Dutch Bouquet', more than 23 organic colours and 126 blocks are used. The few craftsmen are existing who retain this painstaking legacy with great artistry in Rajasthan in natural dyes. Some of the few clusters are Akola, Saganer, Bagru, Barmer, Pipad that even rehearse the art of making this fabric to keep classic heritage alive [26]. In the Akola (District: Chittorgarh, 70 km from Udaipur), the resist is made of bedja (local gum) and oil by scouring (boiling it for several hours). The printing in this region is a specific type, namely, nangna and phetita.

To enable the printer to make correct registration of dabu prints of the block on the cloth, pitch pins are often fixed round the sides of the block, which print tiny dots on the fabric. These pins are engaged to coincide with specific, well-defined points in the thematic patterns, so that impression or repeat joins up on every side design the other prints surrounded by it. Besides blocks, long tables and several sieves along trolleys are required for carrying a printing as mentioned in Fig. 2. The table for dabu block printing is typically made from wood and covered with resilient pads of woollen cloth or several layers of the gunny cloth on which back grey or ledger cloth is fixed. The fabric to be printed is then gummed to fix the cloth (Fig. 2).

The supply sieve to the block consists of the sieve proper made with a rectangular wooden trough with a bottom covered with woollen cloth stretched tightly and secure by nails. A trough is filled with printing paste (thickener) on the surface that swims the sieve trough. The whole assembly is mounted on a trolley which can be wheeled up to move alongside the printing table.

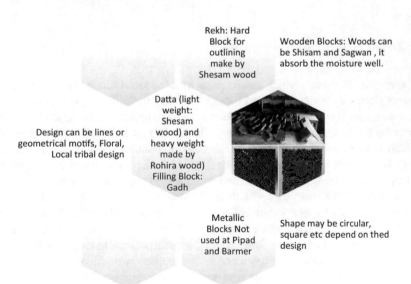

Fig. 2 Type of blocks used in dabu prints [22]

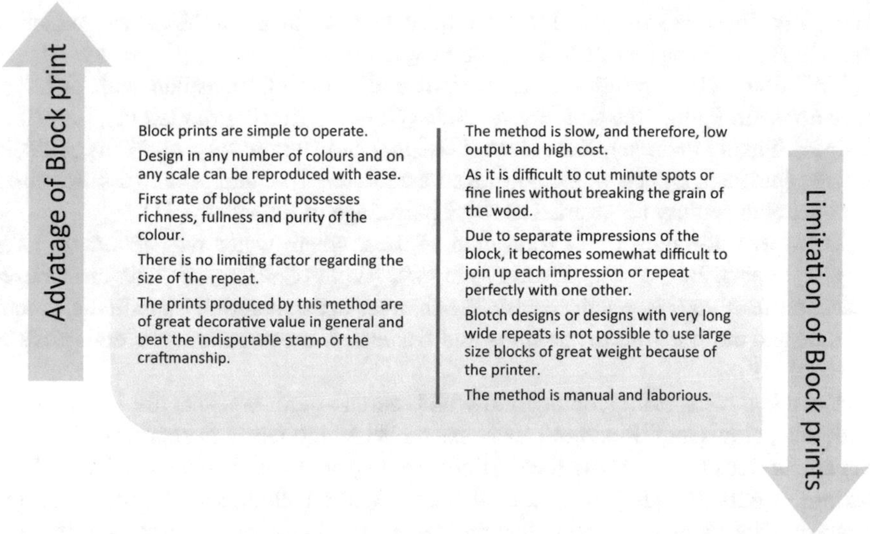

Block prints are simple to operate.

Design in any number of colours and on any scale can be reproduced with ease.

First rate of block print possesses richness, fullness and purity of the colour.

There is no limiting factor regarding the size of the repeat.

The prints produced by this method are of great decorative value in general and beat the indisputable stamp of the craftmanship.

The method is slow, and therefore, low output and high cost.

As it is difficult to cut minute spots or fine lines without breaking the grain of the wood.

Due to separate impressions of the block, it becomes somewhat difficult to join up each impression or repeat perfectly with one other.

Blotch designs or designs with very long wide repeats is not possible to use large size blocks of great weight because of the printer.

The method is manual and laborious.

Fig. 3 Advantage and limitations of block prints

4.1 Procedure

The print paste has flowed on the woollen cloth of the sieve. The Block is carefully placed on this sieve and pressed to pick up a uniform paste layer. It is then stamped on the fabric on the table by giving blows with a small but heavy mallet to ensure a darker impression. An individual sieve is required for each shade at the other swimming tub repeatedly. With the repeated stamping of the colour paste, the pattern is built up, and the process is separated until the organic colours are separately applied, and the whole piece of the cloth is printed. It is then removed, dried in daylight and provided suitable after treatment. Any design in number of colours in any scale can be produce by block printing. There are inherent advantage and disadvantage of block printing process as mentioned in Fig. 3.

4.2 Dabu Prints

The desert state of Rajasthan is known for its striking, kaleidoscopic dressing, produced by distinctive colouring methods (dyeing, printing using vegetable colours) mostly on absorbent and resilient cellulosic fabric. These centres practised the traditional (folk) form of printing usually known as 'Daboo/*DABU*' [18]. The eighth century AD sample of *these prints* found in Asia(central) is believed to be printed in Indian subcontinent [27]. It is denoted by the oldest sample of printed textiles in region. These are supposed to have been finished in India. *'Uchcho',* which is thought fabric prints resist textiles called Dabu [16].

Traditional block printing is an ancient craft form of Rajasthan with different variations throughout the state from Akola (Chittorgarh), Bagru (Jaipur), Sanganer (Jaipur), Pipad (Jodhpur), Kala Dera (Udaipur) and Barmer city, etc. This princely state Rajputana has been well-known for its coloured dyed and printed textiles since the twentieth century blossomed as royal patronage in Rajasthan [21].

Dabu was prevalent in all Rajasthan centres, where water was abundant. Like Bagru (Jaipur), Barmer and Pipad (Jodhpur), Akola (Chittorgarh), etc. are centres where the technique is practised with few changes in the material, ingredients, colour scheme and motifs. *Kalidar,* D*olidar* and *Gwar wali* are popular Dabu end prints in the state [19].

Dabu is a resist print technique in which certain cloth portions are impregnated with wax, clay, Gum, Resin or other resists. When the fabric is successively dyed, the colour does not penetrate the portions impregnated with resistance. The desired pattern is obtained [14]. It is one of the ancient's techniques of printing in the country. Though many printing techniques are employed in this province, it is the most admired and favoured print of the desert state of Rajasthan [15].

5   Status of Dabu Print

It is the soul of India, and the crafts that make our country stand out in the world. It (handprints) is such a craft that can be employed for creating each segment of the fabric design amazing and distinct which are not achievable from machine prints where design and motifs are identical in entire prints. The traditional and exotic hand block printing of Pipad and Barmer has been handed down generations but with the inclination towards industry, traditional printers are remaining fewer in number. It is difficult for block printers to compete against screen print but still Pipad, Balotra and Barmer craft persons are doing this traditional print to keep alive the ancient and fading art of 'Dabu hand block print'. The global awareness and recognition has renewed interest in the use of eco-friendly dyes. There is no doubt that factory prints fabrics are more colourfast and cost-friendly but these prints reflect human efforts and sentiments of craft man.

6   Cluster and Community

This block print form was originated 450 years back when some societies of Chippas (Khatri community) were migrated from Sindh (Erstwhile British India) settled to western Rajasthan and nearby the Luni and Sukdi riverside like any other nomadic settlement.

Khatri believes their origin is from Brahmkshatriya. They link themselves with Brahma [22]. The Sindh region is the source of these prints. The craft persons used to coat the cloth with clay obtained from the pond or riverside and then immersed it in water (turmeric) for a sandy coloured ground. These mud print on the cloth in earthy shades using natural dyes.

At present one family in Barmer and two families in Pipad (Jodhpur region) employ natural dyes and others are operating on napthol-based dyes as per the order and demand of the market. The passion, interest in the craft established the good quality work and hold the particular marketplace. The hard work of these craftsmen is regaining this art to former nobility.

7 Types of Dabu

The impressive resist paste 'Dabu' is mud that acts as natural material made by black clay of nearby pond, lime and gum. Lime water inhibits the cracking of black clay at the printed pattern portion. It also develops the bond with the fabric. The special resist paste (Black clay–lime–gum–insect-eaten wheat mixture or Bidhan), a speciality of conventional printing of state, is commonly known as '*Dabu*'. Dabu performs as a mechanical resist and stops the penetration of colour. The printing paste is differ as per ingredients used in dabu printing paste as meintioned in Fig. 4. This technique is used only for creating patterns with indigo blue. No sharp patterns are achieved due to heavy and sticky printing paste. It is applied with a wooden block on the fabric, and sawdust is sprinkled over it (Fig. 4).

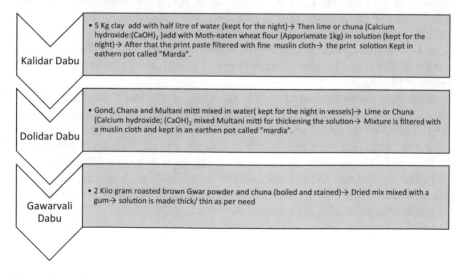

Kalidar Dabu
- 5 Kg clay add with half litre of water (kept for the night)→ Then lime or chuna (Calcium hydroxide:$(CaOH)_2$)add with Moth-eaten wheat flour (Apporixmate 1kg) in solution (kept for the night)→ After that the print paste filtered with fine muslin cloth→ the print solotion Kept in eathern pot called "Marda".

Dolidar Dabu
- Gond, Chana and Multani mitti mixed in water(kept for the night in vessels)→ Lime or Chuna (Calcium hydroxide; $(CaOH)_2$ mixed Multani mitti for thickening the solution→ Mixture is filtered with a muslin cloth and kept in an earthen pot called "mardia".

Gawarvali Dabu
- 2 Kiio gram roasted brown Gwar powder and chuna (boiled and stained)→ Dried mix mixed with a gum→ solution is made thick/ thin as per need

Fig. 4 Type of dabu as per different ingredients [19]

8   Results and Considerations

All the defendants belonged to *the Hindus and Muslims Khatri community*, and their language was *Marwari*. The artisans resided in a nuclear family set up, and all possessed their own *Pakka* houses. Three defendants had undergone secondary schooling, and one had experienced masters at Pipad. In Barmer, the craftsmen are senior secondary. Teenagers have the choice to continue studying or learn the hereditary skill. Most of them are comprehending both, and most of them practised it. The printers work for 8 h every day while the helpers work 8–10 h. The daily production capacity of the Ranamal unit is 400-m running cloth in Barmer.

The herdiatary skill is handed over from one age group to the other by transfer knowledge. All the printers had a ownership of business by an individual. It was heeded that the art of printing functioned as a substantial income source for them. The females also acquired a small portion of finances by bringing out other material craftworks like aari tari, *bandhej with surface* embroidery. The fabric made on a loom has lower width so it is stitched from the middle. The justification for this was, earlier, the fabrics were woven into small widths, and hence to produce a moderate width of fabric, small pieces were stitched together. The respondent mentioned that Rana Mal Khatri, ancestors of Barmer, had come from Western Pakistan (Sindh Region) to this part state.

8.1   Equipments and Tools

Traditinally all tools and equiments used in the prints are handmade. The wood is the main raw material and mud is used by nearby pond. The wood is used from Rohida (Tecomella undulata). After completing their day-to-day work, women made these tools at home. The tools like print blocks (wooden), Stick (stirrer), wooden stand *(Ghode),* table (for prints), *budho* (wooden beater), etc. were made of timber as mentioned in Fig. 5.

In contrast, the equipment like *matka* (Indigo clay dye tank), *Tari* (tray), Kunda (tank for *harda* treatment), *Tari* (tray), receptacles were made from mud of nearby pond. The other types of equipment, like *rangchul* (dyeing tank), were made using cow dung and local bricks. Aluminium was used to make *Charu* (pot in the *rangchul*). The bamboo stick is used for making lower layer (chapari) of tray (tari). The local black mud and salty water play a major role as the colour of these prints are very bright (Chatak rang). The main principle raw material is wood remains for printing tables, blocks, *danda, ghodi* and *budho* (Fig. 5).

Fig. 5
Diffrent tools utilised for
making dabu hand block
from wood [22] (Courtesy:
Manoj Block Print, Barmer,
Rajasthan)

8.2 Ingredients

Several ingredients were used during the whole process of *Dabu* printing as shown in Fig. 6. Block makers carve the blocks, the land gives soil and the ponds awards water. These elements are mixed with the following ingredients.

Many of the ingredients that were used earlier find no use today in the process of production, like *jambe ka tel* (*Eruca sativa* Mill. Seed oil), *Gissi* (camel dung), Ricinus communis Linn. (*erandi ka tel*; castor oil), *leema* (dried lemons), etc.

The well-chosen hues of colour can be changed by treating with mordents. The extraction effectiveness of colourants present in flora and fauna (Animal/Plant/Mineral) sources depends on the media type (aqueous/organic solvent/ or acids or alkali), pH of the dyebath, extraction conditions like bath temperature, time, size of particles and material to liquor ratio (Fig. 6).

Indigoid dyes or Neel: This is maybe one of the important group of natural dyes, acquired from Indigofera tinctoria plant leaves which offers blue colour. The printer bought dyes in the paste from the suppliers from Pondicherry and Andhra Pradesh.

Mordents: The natural colour fixes on the fabric using mordents. There are three varieties of mordents such as metallic (metal salts of aluminium, chromium, iron, tin or copper), tannic acid and oil type of mordents.

Fig. 6 Ingredients used in dabu block printing process (Courtesy: Yaseen Blocks prints, Pipad, Jodhpur)

8.3 Fabric

Previously, the fabric was the source from a nearby area like Pali (Umaid Mill), Kishangarh and Beawar from Rajasthan by regional block printers. In the mid-nineties, most of the mills were getting closed after which cloth was procured from Mumbai (Poddar Mills, Binni), Kamleshwar (south India). It was of 20Ne, 30Ne, and 40Ne and 60Ne cotton cambric English cotton count. Currently, the fabric is obtained from Erode, Tirupur (Tamil Nadu) and Bhiwandi, Ichalkaranji, Malegaon (Maharasthra).

9 Printing Method

Dabu printing is time taking process. It includes the various stages as shown in Fig. 7. Each step needs various tools and ingredients. The process of hand block printing is used for all kinds of *Dabu*, but the steps may vary with the motifs and designs. The printing is done with the help of a mordant, a resist or both. Cotton fabric

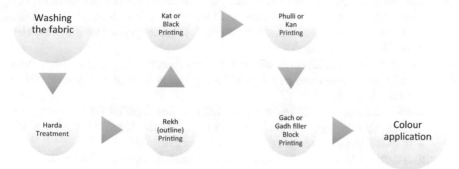

Fig. 7 Various stages of dabu printing

or latha fabric is normally used for printing as per the quality of buyer. Earlier a handspun handwoven fabric was used. Harda treated fabric was printed with double colour blocks in which outline block (Rekh), Iron mordant was used in printing paste and Chrome mordant paste for filling the design (Dutta). The printed cloth is dyed with Dhawari flowers for half an hour (1/2 hour) at boiling temperature. Wash, rinse and dry the cloth, the fabric has beautiful printed design in light and dark brown colours. Alizarin Printing Harda treated fabric was printed with printing paste having alum mordent, alizarin and gum extract in required concentration. Printed fabric was steamed for half an hour, wash, rinse and dry the fabric. The finished sample show the reddish brown colour.

1. Firstly the grey cloth dipped in soda ash and taramira oil (jambe ka tel) solution and then dried in sunlight for 7 days, after that it washed thoroughly on open tank. Presently using hot water/steam make the process is fast and saving the time. This process improves absorbency in the fabrics so fabric is ready for the next process. (Fig. 8).
2. After washing it is soaked into a solution of Harda (Fruit-myrobalan:terminalia chebula) for yellow tanning as shown in Fig. 9.

Fig. 8 Grey cotton cloth and washing tank (Courtesy: Manoj Block Print, Barmer, Rajasthan)

3. The paste is prepared with local babool gum (Acacia gum), multani mitti (fuller's earth), mustard oil, fresh cow dung with water and a sieve it with a fine muslin cloth. The outlines of design where white is required printed with the wooden cloth (Kirana act a resist).
4. The black colour is prepared with the fermentation of Iron, jaggery, Bajra flour (Pearl millet) with water as solvent. The colour is drained and filtered with a fine muslin cloth.
5. Kath—The printing outline is called kath. It develops into dark black as it oxidises by reacting with tannic acid in Harda fabric. Thus, it is then covered with the resist paste with a mixture of Acacia gum and mud.
6. Red Colour: It is made by using tamarind seed, Bajara flour, alum (fitakari). Kirana is an outer design, whereas Kath is inner and space is resisted by mud with Jowar (Sorghum), rice or millet flour and gum of dhawda tree (Anogeissus latifolia). The resist helps in protecting the printed area from Indigo dyes. To prevent the resist from spreading to other areas, fine sawdust (lakdi ka burada) is sprinkled as shown in Fig. 12 on a wet printing surface of fabric. Sawdust has two primary roles first to absorb water from the paste and give additional layers of resistance as shown in Fig. 14.
7. After printing cloth is dried in open sunlight as shown in Fig. 13 and the dried cloth is dipped into an indigo bath which has sodium hydrosulphite that act as a reducing agent. The shade of the fabric depends on several dips and dry operations. The cloth is dipped in Indigo twice which makes blue darker. The fabric is then taken out of the tank, squeezed and opened out to react with the atmospheric oxygen for oxidation of vibrant colours. The oxidation process turns reduced indigo into an oxidised form. For darker tones, the cloth is again dipped in the tank, pulled out and oxidised by air. This process is repeated till the desired dark shade is achieved. The fabric is finally dried flat on the ground under the sun in daylight. Due care is taken while dyeing and drying so it does not get cracked or smashed. The fabric is dried in sunlight in open field or hung by bamboos. That technique is called 'Bichana' or 'Latkana'. Pipad and Barmer both are using indigo dye bath which are 10–15 years old. The hydro is added as per requirement (Fig. 10).

Fig. 9 Dabu block printing at Pipad, Jodhpur, Rajasthan (self print and clicked)

Fig. 10 Indigo dye bath
(Courtesy: Manoj Block
Print, Barmer, Rajasthan)

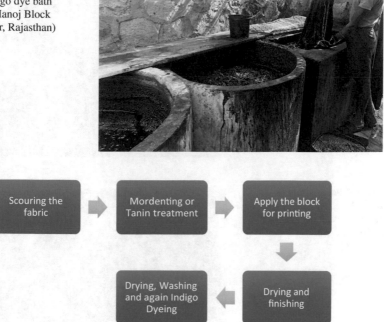

Fig. 11 Flowchart for production for dabu print fabrics

8. The dyed cloth is boiled with alizarin, dhawdi ke phool for one hour thirty minutes in the copper vessel by regular stirring. (Red appears: alum is printed, Indigo: dark blue, Kath: warm black) (Fig. 11).

Previously, the printed fabric was kept underground to treat it in moisture and heat for four months. During this period, the organic colours printed on the cloth get fastened with Alum at underground room (*Bhanwara* 4–6 feet deep) in the earth. Now, this process has been shortened. Now the market demand and to meet big orders the process has been reduced. Now it takes few days to ready the printed cloth and supply it to the clients. The unique designs are provided by the clients like Fabindia, Anokhi, Westside who demands them thousand of metres cloth in one month. Therefore, the customary method of printing and dyeing is no more followed. Sometimes synthetic Indigo and Alzerin have been begun which do not require a long time for preparation (Figs. 12, 13, 14 and 15).

9.1 Application of Dabu on Fabric

The final pattern or design of fabric differs on various processes which occur step by step manner. The area on which mud resist print paste is applied did not absorb

Fig. 12 Dabu block printing
(Courtesy: Yaseen Block
Print, Pipad, Jodhpur,
Rajasthan, India)

Fig. 13 Fabric dried in
daylight (Courtesy: Yaseen
Block Print, Pipad, Jodhpur,
Rajasthan, India)

the colour. The wooden powder (sawdust) is applied on the surface of printed fabric
to absorb the moisture on the fabric. It enables quick drying and stops the plopping
of dyes to other areas. The fabric is dipped in indigo dye bath 3–5 times as per
the desired shade. The oxidation of the indigo happens logically by air oxidation.
Further, the fabric is dry and rinsed. The mud resist paste must be dried by exposed
sunlight. In the rainy season, printers are not functioning. After every dyeing, the
fabric is well rinsed to remove the mud and printing ingredients. Ultimately, the
un-dyed portion where the resist has been applied comes out after the washing. The
recipe of '*Dabu'print* paste is kept undisclosed by families who were working in the
cluster (Fig. 16).

9.1.1 Washing

The resist paste or Dabu and superficial dye which is physically attached to the fabric
are removed by washing the fabric. The fabric is soaked in water for three to four
hours. The resist print paste becomes smooth by absorbing the water. The excess dye
and paste are removed by beating the fabric on a large flat slab of stone. The cloth
beating is normally done at influx of the water. The way of crafting dabu print is so
lengthy that after-effects it's worth it. An eternal craft, dabu print encompasses raw

Fig. 14 Craft man sprinkle
saw dust on dabu print
fabrics (Courtesy: Yaseen
Block Print, Pipad, Jodhpur,
Rajasthan, India)

materials like water and mud. For rich and vibrant look, fabric can be dyed more than once that provides a double and triple dabu look by diffrent colours as shown in Table 3.

10 Material

Handmade carved wooden blocks with fine carpentry tools were used for printing. These blocks are made of teak and Rorda (light-weight wood) from Ardu (Ailanthus exceisa). The artisan carved the wooden blocks with iron chisels of different widths and cutting surfaces. To smoothen the grains these blocks are soaked for 10–12 h in oils before printing. Chhippas have a good amount of block collection which is considered as their core wealth. Most of the printing units were equipped with utensils for colour mixing (drums) and traditional boiler (copper vessel mounted by bricks), long length padded printing table, printing paste tray and paste carrying tray and revolving stool. Whereas in a few unit's traditional boilers (copper vessels mounted by bricks), steaming drums were also found.

Fig. 15 Dried in open sun light

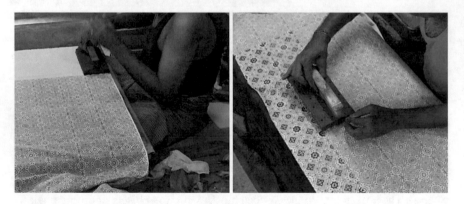

Fig. 16 Resist print (by mud) with Kirana and Khat at Barmer (Courtesy: Manoj Block Print)

11 Motifs

In the sandy desert state of the Rajasthan, colourful prints are picked from nature and its surrounding's elements like birds, human figures, animals, gods and deity

Table 3 Different vibrant colours obtained by dabu Printing

S. No	Organic colours	Colours produce	Process	Mordents used
1	Black dye	Yellow colours. obtained from Harda or *Termimalia chebula (plant fruits)*	The seeds of harda Ist wetted out in cold water for 24 h and the next day converted into a smooth paste with a mortar. 3 Kg of ferrous sulphate boiling with 18 litre of water to preapre the paste.	Ferrous Sulphate (FeSO$_4$)
2	Indigo Dye	Plant *Indigofera Tinctoria*	3 kg Indigo with 500 ltr water add with 300-500 gram lime and hydro (3 Kg)	Stannous chloride with acid, alum or chrome
3	Yellow dye	Flower of *marigold* and *bougainvillea glabara*	3 Kg Pomegranate with 50 gram alum boiling with 18 litre water to make orange colour.	Alum and Tin mordent

are well-known in dabu prints. The sky shade blue, indigo of the evening, twilights red—are customarily seen in the regional clothes. The wavy lines, dots and various geometric patterns are also used with indigo blue shade, Kashish(CuSO$_4$) which grey brown, yellow and red is derived from fruits like pomegranate is the characteristics of this Centre. The traditional motifs printed in Dabu were *pallu, koni* (or *mordi*), *hanso, nagan*, etc. These were printed with wooden blocks called *bhaat*. Motifs used in Dabu prints are generally trivial in size and primarily contain Bel (creepers), Buta and Buti (Floral and Floral net motifs) and Jaal. Conventional patterns of flowers, fruits (Keri- Raw Mango), leaves (Paan ka paata), buds, twisted tendrils and stylized figurative motif of animals (Elephant, Horse, Peacock, Camel), humans (dancing females) along with stylized geometric patterns (Chaupar, Chatai, Kangura) are also the attraction of Dabu printing. When printing started at that time common patterns printed by the local people were Coriander, Chaubundi, Chakri and neem leaves. Some of the traditional motifs used in Pipad printing are Ilaichi buti, Bichu buti, Dhatura buti, Dhania buti, Genda buti, Chaubundi buti, Kamal buti, Gulab buti, Mukut buti, Paan buti, Katar buti, Mirchi buti, Jadhpul buti, Surajmukhi buti, Haathi buti, Ghoda buti, Belpatra ki bel, Neem patti ki bel, Angoor ki bel, Machi ki bel, Kamal phul ki bel, etc. The final outcome of unique designs repeated again and again all over the fabric. Often the black smooth mud paste crashes and seepages that generates veins alike to Batik.

12   Product Range

Generally, Dabu fabric (print) is used to adorn traditional ethnic clothing like the flowing Ghaghra, Odhnis, Chunris, and Safa (Headgear), but it can also be used at home décor products. The favoured dress of the local women, known as 'Fetiya' in the desert state of Rajasthan, was coupled with a bandej lugda. Only a single stitch is used to make the Fetiya. It was prepared by connecting the outer ends of 8–12-m fabric or as long as the customer's interest. The attire was then gorgeously wrapped by Jat, Gadariya and Gujjar females.

Odhnis/Chunris from Jodhpur block print are highly prized. Today, Dabu has found a wide range of made-up garments and home furnishing products. It retains fadat, lungi, angochha, tehmat, dhoti, sarees, whereas accessories like scarfs, rumal (Handkerchief) and headgears are also made with Dabu prints. Not limited to just clothing, Dabu is a favoured category in home furnishings articles like Bhichhauni, table covers, napkins, bed sheets or bedspreads, and quilts are popular furnishing articles crafted from Dabu prints. Traditional Dabu prints were found primarily on absorbent and resilient cotton fabric, but to meet the market's contemporary demands, occasionally cotton and silk blends are also being used. The prints are uniquely designed on cloth made by natural fibres like silk, cotton and fabrics (georgettes, crepes, chiffons, super net sarees). They embrace vibrant natural colours quite well; however, absorbent and resilient cotton fibre continues the most extensively used thread for dabu printing textiles.

13   Maintenance

Dabu print cloth would be first laundered by hand to check colour fastness. After that, the fabric can be machine washed and rinsed with cold water. Dabu print fabric is best air-dried or line dry to avoid fading bright colours in natural Sun-light. Natural dyes usually do not fade, but numerous machine washes might speed up the process of colour fastness (Table 4).

14   Conclusion

The story and folktale of craft are well-known by artists who performing this craft. Mahatma Gandhi mentioned that "The world has enough for man's need but not enough for man's greed!" The hue of cloth inspired by local flora and fauna and distinctive design customs put these hand block prints products separately in today's competitive markets from fast fashion goods. The usage of non-allergic, nontoxic and eco-friendly natural dyes used for printing has become important due to the increasing of environmental attention to avoid some hazards of synthetic dyes.

Table 4 Glossary

S.No	Local names	Meaning
1	Adatiya	Local traders who buy fabric in bulk
2	Angocha	Towel type fabric put on shoulder
3	Babool	Acacia tree
4	Bhanwara	Underground room
5	Bhaat	Wooden block
6	Bichalna	A sprinkling of water against the sun on the printed and dyed cloth spread on the ground
7	Bidhan	Wheat powder
8	Chunna	Lime
9	Datta	Wooden block filled the space
10	Dhawada	Tamarisk small shrub tree (Salt cedar)
11	Dhawde-ke-phool	Tamarisk flowers
12	Fitkari	Alum
13	Fetiya	Ghagra or a long full skirt, often decorated with embroidery, mirrors or bells
14	Gadi	Red colour tome obtained by prints and dyeing
15	Gancha	Hand towel
16	Ghand	Process of boiling the printed, dyed cloth in Alizarin with the mixture of Dhawada flowers
17	Gur	Jaggery
18	Harda	Myrobalan
19	Jowar	Sorghum
20	Kalam	Chisel
21	Lattha	Grey cloth
22	Lugdaa	An extended fabric material draped over the head
23	Kumhaar	Potter

(continued)

However, worldwide, the usage of natural dyes for the colouration of textiles has mainly been confined to artisan/craftsman, small-scale cottage level dyers, printers and small-scale exporters are dealing with high valued eco-friendly textile production and sales. The process has transformed into a significant stretch to fulfil the growing need of the market and save time. The stage remains the same but there is a change in the use of natural ingredients, tools and machines. The production process has not changed in the number of stages, but there is a change in the natural ingredients used at different locations. The synthetic dyes (indigo) replaced the natural indigo to cut down on production time. Wood is replaced by polymer and metal, but conventionally, they were made only with wood. Fortuitously, consumer concern about the unhealthy consequences of chemical pollutants creates a boon for this craft. This segment can

Table 4 (continued)

S.No	Local names	Meaning
24	Lenhga	Skirts
25	Lungi	Lower waist cloth for men
26	Majisht	Madder
27	Sagwan	Commonly called teak scientific name is Tectona grandis
28	Safa	Head gear
29	Rohira	Tecomella undulata (desert teak)
30	Rekh	Outline
31	Pattu	Shawl
32	Mitti	Black clay
33	Methi	Fenugreek
34	Sheesham	Indian rosewood (*Dalbergia Sisoo*)
35	Sakoor	Fruit like Myrobalan earlier used in Sindh in place of harda
36	Tanka	Lager pot made by clay
37	Tapper	Hammering with hands

sightsee and revitalise the application of these products in ready-made garments. They have the resilience to resist the market if formal organisations support and expand the marketing facilities of these products. Earlier, in a *Dabu*, one would find green colour minimal, but today, many yardages are printed in the same colour. Traditionally, K*amarbands (*band of cloth worn around the waist*), lungis* and turbans fabrics were made by dabu printing; nowadays an exclusive range of products like Kurts, Saris, bed sheets*, dress* material is made by these crafts, etc. Due to the legacy of the traditional craft of *Dabu*, it is still as bright as ever and competing in the merchandise to preserve the inheritance of our country.

References

1. Pizzuto JJ et al (2012) Fabric Science. 10th, Edition. Fairchild Books, A division of Condnast publications, New York
2. Vigo TL (1994) Textile Chemical Processing and Its Properties. Elsevier Publication, pp 177–185
3. Schofield JS (1984) Textile printing 1934–1984. Textile Science and Technology (11), pp 112–192
4. Prayag RS (1999) Technology of Textile Printing. Shree J Printers
5. Hossain AH et al (2015) Overview of Piece Printing Process in Textile Industry. IOSR J Polymer Textile Eng (IOSR-JPTE), 2(3):17–28
6. Kasikovic H et al (2016) Textile Printing – Past, Present, Future. Glasnik Hemiĉara, Tehnologa i Ekologa Republike Srpske, pp 35–46, Accessed 07 February, 2019
7. Shenai VA (1985) Technology of Printing: Technology of Textile Processing. (6), Mumbai: Sevak Publication

8. Ganguly D (2012) A Brief study on Block Printing Process in India. Fibre to Fashion, Accessed 11 March 2021
9. Chavan RB (1996) Technological Revolutions in Textile Printing, Indian J Fiber and Textile Res 21:50–56
10. Gokarneshan N (2018) Advances in Textile Printing. International Journal of Textile Science and Engineering,(01), pp. 1–5
11. Rattee ID (1977) Melt-Transfer-and-film-release-systems-of-transfer-printing. Journal of Society Dyers and Colourists, (93), p 190
12. Marsh JT (1950) An Introduction to Textile Finishing. Chapman & Hall Ltd., London
13. Štefko R, Steffeck V (2018) Key Issues in Slow Fashion: Current Challenges and Future Perspectives, Sustainability
14. Jayakar P (1950) Indian Printed Textiles, All India Handicrafts Board, Ministry of Commerce and Industry, New Delhi
15. Ahivasi D (1976) Range Evam Chhapa Vastra. Banaras
16. Chandra M (1973) General of Indian Textile History". Oriental Publishers, Delhi
17. Roy SR (2011) Vegetable Dyes and its Application of Textiles, Dept. of Silpa-Sadhna Visva-Bharti, Birbhum, West Bengal pp. 3
18. Kaur J (2011) DABU - A Unique style of mud printing. Birbhum, Silpa Sadana, VisvaBharati, 2nd – 4th December 2011, pp. 157–163
19. Bhandari V (2004) Costumes, Textiles & Jewellery of India. Mercury Books, New Delhi
20. Karolia A, Buch H (2008) Ajarkh, The Resist printed fabric of Gujarat. Indian Journal of Traditional Knowledge, (IJTI) 7(1):93–97
21. Upadhyay H et al (2016) Bagru: The Traditional Eco-Friendly Hand Block Printing of Rajasthan. International Journal of Textile and Fashion Technology (IJTFT) 6(6):pp 37–44
22. Meena M (2015) Ajrakh Document. Kota,
23. Hada JS (2015) Dyeing with Natural Dyes : A Case Study Of Pipad Village, District Jodhpur, Rajasthan. https://doi.org/10.13140/RG.2.1.5049.3204
24. Vankar PS (2000) Chemistry of Natural dyes. Resonance, 5(10):73–80 https://doi.org/10.1007/BF02836844
25. Gulrajani ML, Gupta D (1992) Natural Dyes and Application to Textiles. Department of Textile Technology, Indian Institute of Technology, New Delhi, India. https://www.scirp.org, accessed 2020
26. Gupta F (2018) Dabu printing - An Ancient Art of Rajasthan. https://www.faridagupta.com/blog/dabu-print-an-ancient-art-of-rajasthan.html. [Accessed 15 October 2021]
27. Dabu Printing (2021) https://Industries.rajasthan.gov.in [Accessed 15 October 2021]

Nanoencapsulation Methods for Herbal-Based Antimicrobial Agents—A Sustainable Approach

M. Gopalakrishnan, R. Prema, and D. Saravanan

Abstract Health and hygiene textiles play an important role in the apparel sectors. Nowadays, various antimicrobial agents are applied to textile materials and most of them are synthetic agents. Though synthetic antimicrobial agents provide relatively good antimicrobial properties, they have issues of eco-friendliness. A significant number of natural herbal-based antimicrobial agents are also applied on textiles, especially cotton materials. These natural product-based antimicrobial agents are applied in different ways to improve the durability and sustainability of the finish. Nanoencapsulation method provides durability and sustainability to the finish when compared with direct methods. In this chapter, various herbal-based antimicrobial agents and their application methods are discussed.

Keywords Antimicrobial · Cotton · Finishing · Herbal · Microencapsulation · Nanoencapsulation

1 Introduction

2 Antimicrobial Agents For Textiles

Antimicrobial agents used in textile industry were discovered previously for some other industries like food, health care, and swimming pool. Minimum Inhibitory Concentration (MIC) values denote the efficiency of antimicrobial agents. Both

M. Gopalakrishnan (✉)
Department of Textile Technology, Bannari Amman Institute of Technology, Sathyamangalam 638 401, India

R. Prema
Department of Electronics and Communication Engineering, Bannari Amman Institute of Technology, Sathyamangalam 638 401, India

D. Saravanan
Department of Textile Technology, Kumaraguru College of Technology, Coimbatore 641 049, India

Agar dilution and broth dilution methods are used to find out the Minimum Inhibitory Concentration (MIC) [1]. The huge number of metals, metal salts, quaternary ammonium compounds, and natural plant extracts are used as antimicrobial agents for textile applications.

3 Medicinal Plants as Antimicrobial Agents

Various plants used in herbalism (herbal medicines) are known as medicinal plants. The word "herb" originates from the Latin word, "*herba*". Herbs come from trees and shrubs. Nowadays, all the parts of the medicinal plants, fruit, flower, stem, bark, stigma, seed, root, or leaf are referred to as herbs. Certain common names, botanical names, parts of the plant used for medicinal purposes, and their botanical family names are given in Table 1. World Health Organization (WHO) recently reported that 80% of people consume herbal medicines worldwide. According to a WHO report [2], over 21,000 plant species have medicinal values.

3.1 Neem (Azadirachta Indica)

The medicinal plant, *Azadirachta indica* (Fig. 1), a versatile medicinal plant, has the ability to protect against most microorganisms. *Meliaceae* has two closely related species, *Azadirachta indica* and *M. azedarach*. *Azadirachta indica* is known as Indian *Neem* (Indian lilac), The evergreen tree, *Neem*, is cultivated all over India and every part of *Neem* including seeds, barks, leaves, twigs, roots exhibits significant medicinal values. The major advantage of *Neem* is that the chemicals present in the *Neem* attack

Table 1 Medicinal plants—parts used and botanical family

S. no	Common name	Botanical name	Part of the plant	Botanical family
1	*Neem (Indian lilac)*	*Azadirachta indica*	Leaf	*Meliaceae*
2	*Tulsi*	*Ocimum sanctum*	Leaf	*Lamiaceae*
3	*Aloe vera*	*Aloe vera*	Leaf	Genus *Aloe*
4	*Pomegranate peel*	*Punica granatum L*	Fruit	*Punicaceae*
5	*Nochi*	*Vitex negundo*	Leaf	*Lamiaceae*
6	*Asian pigeonwings*	*Clitoria ternatea*	Flower	*Fabaceae*
7	*Mexican mint*	*Coleus amboinicus*	Leaf	*Lamiaceae*
8	*Kantal*	*Gloriosa superba*	Flower	*Colchicaceae*
9	*Nutgrass*	*Cyperus rotundus*	Seed	*Cyperaceae*
10	*Kodukkappuli*	*Pithecellobium dulce*	Leaf	*Fabaceae*
11	*Kuppaimeni*	*Acalypha indica*	Leaf	*Euphorbiaceae*

Fig. 1 Azadirachta Indica

only the hormonal system of the insects. All the parts of *Neem* are extensively used to control dwelling pests and crop pests [3–5]. Siddiqui in 1942, reported the isolation of Nimbin from the *Neem* oil and since then more than 135 compounds are identified and isolated from various parts of the *Neem* plant [6]. Nimbidin has several biological activities, extracted from seed oil. The other compounds such as Nimbinin, Nimbin, Nimbidin, Nimbidic acid, and Nimbolide have also been extracted from crude seed oil [7]. Certain compounds, isolated from *Neem*, have bioactive properties; Nimbidin and Nimbin have the ability to inhibit the growth of Mycobacterium tuberculosis and thus has bactericidal properties.

Neem extracts are extensively applied on cotton fabrics [8–10], cotton/polyester blended fabrics [11], silk fabric [12], and wool fabrics [13], by direct applications, either pad-dry-cure or exhaust methods. The *Neem* extracts exhibit good antibacterial activities against *Staphylococcus aureus* and *Escherichia coli*. The direct method of finishing showed very poor antimicrobial activity. Whereas, the microencapsulation method exhibits good wash durability up to 15 wash cycles [14], and nanoencapsulation method extents up to 25 wash cycles [15–17], whereas the plasma treatment, prior to the antimicrobial finishing, enhances efficacy of the antimicrobial activity significantly [18].

3.2 Tulsi (Ocimum Sanctum)

Ocimum sanctum (Fig. 2), commonly known as *Tulsi* or holy basil, is a perennial plant of the family *Lamiaceae*, cultivated in the Indian subcontinent and Southeast Asian tropics. The leaves of tulsi give aroma and the extracts protect the human from mosquito bites and provide relief from headaches. *Tulsi* is commonly used in Ayurveda and other traditional medicines for the essential oils present in *Tulsi*. *Ocimum sanctum* leaves contain volatile oils (0.7%) and 71% of the volatile oil present in *Ocimum sanctum* is eugenol and 20% of the volatile oil present in *Ocimum sanctum* is methyl eugenol.

Fig. 2 Ocimum sanctum

Various compounds including ursolic acid, carvacrol, sesquiterpene hydro-carbon caryophyllene, apigenin, apigenin-7-0-glucuronide, luteolin, luteolin-7-0-glucuronide, molludistin and orientin have been isolated from the *Ocimum sanctum* leaves. The flavonoids vicenin and orientin are isolated from the *Ocimum sanctum* leaves besides various phenolic compounds. Traces of zinc, sodium and manganese are also found in the leaves.

Joshi et al. [19] analysed the antimicrobial activity of medicinal plants against the most pathogenic microbes and found that the natural antimicrobial agents are more inhibitory against the Gram-positive microbes than Gram-negative microbes [19]. Aqueous extracts of *Ocimum sanctum* exhibit insecticidal and antibacterial activity against most bacteria. But *Ocimum sanctum* does not exhibit antimicrobial activity against *E. coli*, *Shigella*, *Salmonella*, *Staphylococcus citreous*, and *Aspergillus niger* [20].

Eugenol (4-allyl-2-methoxy-phenol) is the principal component in *Tulsi* extracts and eugenol has the potential to inhibit the production of aflatoxin in the food industry. Ethanolic extracts of *Tulsi* show relatively better results than *Azadirachta indica* *(Neem)* in an antiviral agent. Combinations of *Ocimum sanctum* and *Cassia alata* exhibits better anti-cryptococcus activity [21].

Ocimum sanctum extracts were applied on cotton materials by the direct method, either by exhaust method [9] or pad-dry-cure method, or microencapsulation method [22]. The antimicrobial activities of the direct and microencapsulation methods exhibit good results with low durability, whereas microencapsulation shows better activities even after 15 wash cycles [22].

3.3 Aloe Vera

Leaves of *Aloe vera* (Fig. 3) used in cosmetic and wound healing products contain more than 75 nutrients, 20 minerals, 12 vitamins and 18 amino acids. Totally 200 active ingredients have been identified in *Aloe vera* leaves and have good inflammation properties. Therefore, *Aloe vera*-based textile products are used in wound dressing and sutures [23].

Ammayappan and Jeyakodi Moses reported *Aloe vera* gel to be normally applied alone and in combinations with chitosan and curcumin on cotton, wool, and rabbit hairs. The antimicrobial activity of *Aloe vera* is better than the natural antimicrobial agents chitosan and curcumin. The combination of *Aloe vera* with chitosan and curcumin improves the activity [24].

Jasso et al. reported different polysaccharides, derived from *Aloe vera* including acetylated glucomannan, glucomannan with different molecular weights, galactoglucomannan with different compositions, galactogalacturan and acemannan. The long-chain polymer of acemannan—randomly acetylated, has immunomodulation, antifungal, antibacterial, and antitumor properties [25]. Tannin compounds, in *Aloe vera,* change the colour to brownish-green colour. Arunkumar and Muthuselvam reported the maximum antimicrobial activity of *Aloe vera* from the acetone extract than the aqueous extract [26].

Aloe vera extracts of 5, 10, 15, 20, and 25 GPL were applied on cotton fabrics with glyoxal as the cross-linking agent and displayed good antimicrobial activity than untreated samples [27]. *Aloe vera* gel is applied on cotton fabrics along with a cross-linking agent 1,2,3,4-butanetetracarboxylic acid using the pad-dry-cure method in which the ingredients present in *Aloe vera* gel are attached to the hydroxyl groups of cotton cellulose [28]. Another group of researchers reported the durable antimicrobial activity using *Aloe vera* gel finished on cotton fabrics [23, 29].

Fig. 3 *Aloe vera* plant

Fig. 4 *Pomegranate* peel

3.4   Pomegranate Peel (Punica Granatum L)

In many papers, the antioxidant and anti-inflammatory properties of *Pomegranate* peel (Fig. 4) are reported [30]. The flavonoids and tannins are responsible for antioxidant properties. *Pomegranate* extracts were tested against a number of bacteria and the *Pomegranate* extracts show activity against all microbes [31]. Wool is dyed with natural dyes after mordanting with tannin-rich *Punica granatum L.* The wool fabrics pretreated with *Punica granatum L* at 5% concentration show a significant increase in antibacterial activity [32]. *Punica granatum L* can be applied on cotton by the direct exhaust method, which shows good activity [33].

3.5   Vitex Negundo (Nochi)

Vitex negundo (Fig. 5) is a 3 m height of aromatic shrub that belongs to a Verbenaceae family [34]. The leaves have anti-inflammatory activity, antihistamine properties, and analgesic properties. Polyphenolic compounds in *Vitex negundo* are responsible for the antioxidant properties of the herbal [35]. The plant has many polyphenolic compounds, glycosidic iridoids, terpenoids, and alkaloids [36].

Many phytochemicals, Hexanoic acid, 4H-Pyran-4-one, ethyl ester, 2,3-dihydro-3,5-dihydroxy-6- methyl-, ethyl ester, Hexadecanoic acid, Caryophyllene, 3-hydroxy-, Ledol, Benzoic acid, Aromadendrene oxide, Octadecatrienoic acid, Phytol, and Vitamin E contributes the antioxidant and antimicrobial properties. The phytocomponents which have antimicrobial properties present in *Vitex negundo* leaf are listed in Table 2 [37].

Extracts of *Vitex negundo* leaves exhibit the potential to be repellent against various adult vector mosquitoes [38] and flavones isolated from *Vitex negundo* show the ability to inhibit microbes [39], and also possess antivenom activities against snake poison [40].

Fig. 5 Vitex negundo

Table 2 Phyto-components and activities of *Vitex negundo* leaves

S. no	Name of the compound	Nature of the compound	Activity
1	4H-Pyran-4-one, 2,3-dihydro-3,5-dihydroxy-6-methyl-	Flavonoid fraction	Antimicrobial, anti-inflammatory
2	Caryophyllene	Sesquiterpene	Anti-tumour, analgesic, antibacterial, anti-inflammatory, sedative, fungicide
3	Benzoic acid, 3-hydroxy-	Benzoic acid compound	antibacterial
4	Ledol	Sesquiterpene alcohol	Antimicrobial, anti-inflammatory
5	Aromadendrene oxide	Sesquiterpene oxide	Anti-tumour, analgesic, antibacterial, anti-inflammatory, sedative, fungicide
6	Phytol	Diterpene	Antimicrobial, anticancer, anti-inflammatory, diuretic

Vitex negundo is used to synthesize silver nanoparticles, as an alternative method to the chemical synthesis of Ag/NPs which possess significant antibacterial activities against both Gram-positive and Gram-negative bacteria.

Vitex negundo was applied on cotton fabric by exhaustion and microencapsulation methods and both methods exhibit good protection against a wide range of bacteria. The bactericidal effect is developed in the exhaustion method, while the microcapsules display bacteriostatic effect [41].

3.6   Clitoria Ternatea Linn

Clitoria ternatea Linn (Fig. 6) is a perennial herbaceous plant that belongs to the family Fabaceae and it is the native of tropical Asia, Indonesia, and Malaysia. It is also known as Butterfly pea and in Tamil language, known as Kakkattan/Sangu-poo. Hexacosanol, anthoxanthin, taraxerol, lactones, phenol glycoside, aparajitin, phydroxycinnamic acid polypeptide, alkaloid, stigmast-4-ene 3, kaempferol, 6-dionie, cyanine chloride, linolcic, clitorin, palmitic, tannins, resins, oleic, stearic, linolenic acids, finotin, etc. are the main constituents of *Clitoria ternatea Linn.*

The leaves, seeds, and roots of *Clitoria ternatea Linn.* are used in "Ayurveda" and in "Medhya Rasayana" the traditional Indian medicine, as a restoring recipe to boost the memory, intelligence, and for the treatment of neurological disorders [42].

The extracts of seeds, callus, and leaves of *Clitoria ternatea Linn* display good antimicrobial activity against the fish pathogenic microbes [43]. Significant antimicrobial activity of the *Clitoria ternatea Linn* leaves was reported by Anand et al., extracted with different solvents, against the most pathogenic microbes and the methanol extracts exhibits more prohibitory impacts than other solvents [44].

Clitora teratea extracts are used for the synthesis of silver nanoparticles and the antimicrobial activities of such silver nanoparticles derived from root extracts [45] and the whole plant extracts [46] have been studied and positive results of the antimicrobial activity have been reported.

Fig. 6 Clitoria ternatea Linn

Fig. 7 Coleus amboinicus

3.7   Coleus Amboinicus

Coleus amboinicus, well known as Indian borage (Fig. 7), semi-woody, perennial, subshrub, belongs to the family of *Lamiaceae* [47]. Secondary metabolites, phenolics, and flavonoids that have the ability to prevent the free radical scavengers are present in leaf, root, and bark of the *Coleus amboinicus* and ensure antimicrobial, antioxidant, and anti-inflammatory properties [48]. Chronic cough, fever, and headache are treated with the extracts of *Coleus amboinicus.* It is observed that the antioxidant properties are less in aqueous extracts than in ethanolic extracts and the ethanolic extracts of *Coleus amboinicus* have the potential to prevent the growth of Gram-positive and Gram-negative microbes [47, 49].

The extracts obtained from hexane solvent and freeze-dried samples did not show any activity against these disease-causing microbes [50]. The lowest minimum inhibitory concentration value is obtained with acetone extracts, while the antimicrobial activity of the samples extracted with ethyl acetate showed relatively better activities [51]. The phenolic compound present in ethyl acetate and acetone are responsible for the increased antimicrobial activity and antioxidant properties [52].

The size silver nanoparticles synthesized from *Coleus amboinicus* lour ranges between 4.6 and 55.1 nm [53].

3.8   Gloriosa Superba

Gloriosa superba, a semi-woody perennial climber plant (Fig. 8), belongs to the Liliaceae family, a native of tropical Africa and found throughout South Asian countries [54, 55]. *G. Lutea, G. Planti, G. Sudanica, G. Superba, G. Grandiflora, G. Simplex, G. Rotheschildiana,* and *G. Longifolia* are some of the species associated

Fig. 8 *Gloriosa superba*
(plant and seeds)

with *Gloriosa superba* and have hollow stems of 6 *m* height. All the parts of *Gloriosa superba* have medicinal values (Jana and Shekhawat 2011).

Flowers of *G. Superba* appear a wavy-edged, brilliant yellow and red in colour. Numerous seeds are found in the pods of *Gloriosa superba* with rounded in shape [56]. Tuberous root oil cures the joint pains caused by arthritis. Leaves of *G. Superba* potentially cure ulcers, expel the placenta, haemorrhoids, and the seeds are a potential cure against cancerous growth.

3.9   Cyperus Rotundus

Cyperus rotundus, a nutgrass, is a perennial plant of the Cyperaceae family (Fig. 9). The plant grows up to 140 cm in height, a native of Africa, but it is distributed and grow in the tropical and subtropical region [57]. *Cyperus rotundus* is resulting from its tubers and in Indian folk medicines, rhizomes of *Cyperus rotundus* were used to cure fever, pains, nausea, dysentery, and various blood disorders [34]. *Cyperus rotundus* contains alkaloids, oils, glycosides, flavonoids, saponins, tannins, carbohydrates, starch, and proteins with the traces of Mg, Cr, V, Co, and Mn [57]. The antimicrobial activity of the *Cyperus rotundus* is reported by Parekh and Chanda against various microorganisms. Methanolic extracts display better antimicrobial activities when compared to aqueous extracts [58].

3.10   Pithecellobium Dulce

Pithecellobium dulce (Fig. 10) is a tree of the family Fabaceae, a native of the Pacific

Fig. 9 *Cyperus rotundus* and nuts

Fig. 10 Pithecellobium dulce

coast and Mexico. It can grow up to 10–15 m and the fragrant, sessile, and greenish-white flowers can reach 12 cm in length. Due to coiling, the flowers appear shorter in length. The shell turns pink and opens an edible pulp, once it is ripe.

Pithecellobium dulce possesses anti-inflammatory, antivenom, and antimicrobial properties [59]. Leaf extracts of *Pithecellobium dulce* with distilled water, acetone, chloroform, and methanol exhibit antimicrobial activities against 20 different microorganisms, exhibiting variable zones of inhibition ranging from 7 to 27 mm. Extracts made from solvents are more protective than aqueous media. Tannins suppress the activity of various bacteria, fungi, yeast, and virus [60]. The antimicrobial activity of the extracts is realized by the presence of anthraquinones, flavonoids, alkaloids, proteins, tannins, cardiac glycosides, sugars, and terpenoids [61].

3.11 Acalypha Indica

Acalypha indica (Fig. 11) commonly known as Indian *Acalypha*, the family of *Euphorbiaceae*, found in Asian countries including India, Sri Lanka, Pakistan, Yemen, and tropical South America and Africa. *Acalypha indica* commonly grows in waste places or fields and grow up to 80 cm in height [62, 63]. Upper parts of the plants and roots are used as medicines for pneumonia and asthma, and the extract has the antimicrobial activity. Flavonoids, Tannins, acalyphamide, cyanogenic glucoside acalyphin, succinimide, aurantiamide, and pyranoquinolinone alkaloid are reported from the extracts of *Acalypha indica* [64].

Antimicrobial and antifungal activities of *Acalypha indica* were tested against Gram-positive and Gram-negative microbes by disc diffusion method. Aqueous, ethyl acetate, hexane, chloroform, methanol, and ethanol extracts were used and the results show that all the extracts possess antimicrobial activities against many

Fig. 11 Acalypha indica

pathogen-microbes, whereas the extracts made from chloroform is significant in terms of antifungal properties [65].

3.12 Other Medicinal Plants

Other medicinal plants that possess antimicrobial activity are tea tree, azuki beans, lemongrass, turmeric, eucalyptus oil, onion skin, clove oil, *Diospyros mespiliformis* [23, 66], which are used on finish cotton fabrics for antimicrobial activity.

4 Application Methods

Herbal extracts are applied to textile materials by different application methods to impart various finishes and value addition. In this section, direct exhaust, direct padding, microencapsulation, and nanoencapsulation methods are commonly used to finish the textile materials that are described.

4.1 Direct Methods

Direct method is the easiest method to apply the active ingredients. In this method, no additional treatments are needed to finish the active ingredients. Exhaust and padding methods are commonly used in textile finishing. In the exhaust method, the active ingredients are dissolved in a suitable medium and the bath is set up with the required material-to-liquor ratio. Normally, the material-to-liquor ratio is high in the case of exhaust methods, whereas in the padding method, the active ingredients are dissolved with a very low material-to-liquor ratio and applied on material by padding using mangles with pressure.

4.2 Microencapsulation Method

The surface area of the microparticles is much higher than the normal particles that reducing the number of active ingredients and the thickness of the coating. Microencapsulation is an encapsulation of an active ingredient at microscales. This microencapsulation technique was earlier used in carbonless copy-paper. Now, this technique is widely used in almost all industrial applications including textile finishing and products. A wide variety of materials (core) has been encapsulated with polymeric materials (capsule) with different techniques, described below.

4.3 Emulsification–Solvent Removal Processes

In this method, polymers are miscible with volatile organic solvents, chloroform or dichloromethane (DCM). The extracts are dispersed in a polymer solution and emulsified in an aqueous solution. Then the solvent is removed by evaporation or coagulation process to separate the microcapsules droplets from the emulsion at reduced or atmospheric pressure. The resulting microencapsulated particles are washed, filtered, centrifuged, and lyophilized. In this process, the morphology and size of the microspheres are influenced by the solvent removal process [67]. If the core ingredients are volatile or have affinity towards the organic phase, the removal during extraction, the large volume of water is used. The characteristics of the microstructures are based on the rate of solvent removal and porous microspheres that are formed when the solvents are removed at a faster rate [68].

4.4 Organic Phase Separation (Coacervation)

Coacervation is the formation of the coacervate, the process of separation of colloids from a solution rich in colloid, and the remaining solution is low in colloid content [69]. Unique characteristics are that coacervate forms a separate layer in the solution, and droplets are formed in the polymer-poor phase. Coacervates are concentrated, viscous, and have the ability to bind solids or liquids [70].

4.5 Spray Drying

In spray drying, polymer solutions containing the active ingredients are sprayed out into a desiccating chamber, using compressed nitrogen or air. Microparticles are dried using the stream of warm air [71–74] and this process involves three steps: (i) aerosol formation, (ii) drying of aerosol using warm air, and (iii) separation of dried aerosol.

The characteristics of the microcapsules or microspheres are based on the initial formulation whether it is in the form of a solution, suspension or emulsion. The biodegradable polymers PLA, PLGA, PCL are successfully microencapsulated by this method. Bromocriptine microspheres were the first product to be produced in the commercial method by spray drying [75].

4.6 Solvent Free Techniques

Preparation of microspheres in phase separation, emulsification, and spray drying methods involves organic solvents like DCM, ethyl acetate, or ethyl formate, and chloroform to dissolve the polymers such as PLGA, PCL, and PLA. But sustainable production of microcapsules involves biodegradable polymers.

4.7 Supercritical Fluids (SF)

This method of production of microcapsules contains very less amount of residual organic solvents. Several researchers use supercritical fluids as drug delivery systems [76, 77]. The basic techniques to develop microcapsules using supercritical fluids are: (i) supercritical fluid as a solvent for stabilizing the active ingredient, (ii) supercritical fluid as an anti-solvent—induce precipitation of active core molecule or a carrier when brought into contact with a solvent, and (iii) act as a spray enhancer.

Controlling viscosity and interfacial tension of the supercritical fluid can regulate the size of the droplets sprayed. Due to the great mass transfer rates of the supercritical fluids, SF is best suited for spray-based processes, moreover separation of supercritical fluids from the capsules is easy and eliminates the large quantities of solvent as the by-product [78]. Low critical pressure and temperature, non-flammable nontoxicity, recyclable, inexpensive and safe nature, carbon dioxide are used in pharmaceutical applications as a supercritical fluid.

4.8 Dry Processes—Milling and Grinding

Milling and grinding are other processes used to produce microcapsules without residual organic solvents. This was first developed by Boswell and Scribner in 1973, who milled the polymer mixtures after frozen into fine particles, ranging from 1 to 200 μm. The resultant microcapsules were used for drug delivery systems [79].

4.9 Nanoencapsulation

Nanoparticles are solid in nature, with diameter ranging from 1 to 100 nm in size, and may or may not be biodegradable. The term "nanoparticle" is the generic term used for nanospheres and nanocapsules. In nanospheres, the drugs are adsorbed over the surface of the spheres or they may be encapsulated within the spheres. In nanocapsules, the drug is limited in vesicular systems covered by a polymeric membrane [80]. The method of production of nanoparticles is categorized into two main classes—(1) polymerization method and (2) direct from macromolecules.

4.10 Polymerisation Methods

Emulsion polymerization and interfacial polymerization methods are used to produce nanoparticles. In emulsion polymerization, the organic and aqueous medium is used as the solvent.

4.11 Emulsion Polymerisation Method

Emulsion polymerization is the fastest method to produce nanoparticles [81]. There are two methodologies: continuous organic phase and an aqueous phase. In the former method, monomers are dispersed into an emulsion or into non-soluble material for monomer or inverse microemulsion. Nanospheres of polyacrylamide were first developed by this method [82, 83]. Aggregation of nanoparticles is protected using surfactants or soluble protective polymer [84]. This procedure requires, toxic organic solvents, monomers, initiators, and surfactants and moreover, the process is difficult to produce nanoparticles.

Nanoparticles, polyethyl cyanoacrylate (PECA), polymethyl methacrylate (PMMA), and polybutyl cyanoacrylate were developed by dispersion of monomers into solvents such as n-pentane, toluene, and cyclohexane. Fluorescein, Triamcinolone, timolol, and pilocarpine are encapsulated by this method [85].

In aqueous phase method, monomers are dissolved in an aqueous continuous phase solution with an initiator, which may be ionic or a free radical. Alternatively, using high energy radiation, ultraviolet or strong visible light or g-radiation, the monomer molecules are transformed into an initiating free radical [86].

4.12 Interfacial Polymerization

Nanoencapsules of poly(alkyl cyanoacrylate) are prepared with this method, which produces the nanoencapsules at a faster rate due to the rapid polymerization [87]. The polymerization is initiated by the ions present in the medium [88]. Cyanoacrylate with the suitable drug was dispersed in a mixture of oil-and-ethanol. The mixture is then transferred into the aqueous solution slowly with or without surfactants. Nanocapsules are developed spontaneously by polymerization of cyanoacrylates. Polymerization initiates once the cyanoacrylates contact with ions present in water. Then the nanocapsules are purified by evaporating the solution under vacuum. The nanoencapsules of poly (isohexylcyanoacrylate) and poly (isobutylcyanoacrylate) are prepared by this method [89].

4.13 Interfacial Polycondensation

Interfacial polycondensation is another method to prepare polymeric nanoparticles by interfacial polycondensation of lipophilic and hydrophilic monomers with or without surfactants. Nanocapsules produced by this method are smaller than 500 nm. Polyurethane and poly(ether urethane) copolymers are synthesized by a modified interfacial polycondensation method [90, 91].

4.14 Nanoparticles from Preformed Polymers

The non-biodegradation or slow biodegradation of monomers are used in micellar polymerization. The toxic nature of the residual molecules in the polymerization medium, forced the use of preformed polymers instead of monomers.

4.15 Emulsification/Solvent Evaporation

Two steps are involved in the emulsification-solvent evaporation method. In the first step, the emulsification in an aqueous phase takes place and in the second step, the evaporation of polymer solvent takes place resulting the precipitation of nanospheres. The dissolved drug-containing polymer solution is dispersed into nanodroplets with dispersing agents and chloroform, a non-solvent. The drug of the nanosphere precipitates out in the form of a fine dispersion. The residual solvents are removed subsequently by raising the temperature. The size of the nanosphere is depended on the stirring rate, viscosity of polymer solution, type and quantity of dispersing agent, and temperature [92].

4.16 Solvent Displacement and Interfacial Deposition

It is a similar method based on spontaneous emulsification of the organic internal phase containing the dissolved polymer. This method produces only nanocapsules, whereas the former method produces both nanosphere and nanocapsules [93, 94].

4.17 Emulsification/Solvent Diffusion

The polymer which is encapsulating the core is dissolved in a partially water-soluble solvent and emulsified in the solution containing a stabilizer. Production of nanoparticles diffuses the solvent of the dispersed phase by the excess of water when the organic solvent is partially miscible with water. The resultant nanospheres or nanocapsules are obtained after removing the organic solvent [94].

4.18 Salting-Out with Synthetic Polymers

In this modified emulsification/solvent diffusion method, the solvent, which is miscible in the aqueous medium, is separated out by the salting-out technique. In this method, the solvent is used to dissolve the polymer and drug, and then subsequently emulsified in the aqueous solution containing an electrolyte such as calcium chloride, magnesium chloride, and magnesium acetate. This diffusion of solvent induces the formation of nanospheres [94, 95].

Preparation of albumin nanospheres involves two methods, the first one is thermal treatment at elevated temperatures normally at 95–170 °C and another one is chemical treatment with iso-octane, vegetable oils, or aqueous medium. The other methods are either of these two methods with slight modifications [96]. In the thermal method, the albumin droplets are stabilized at 175–180 °C for 10 min and to reduce the viscosity of the oil phase, the mixture is diluted with ethyl ether after cooling. This method is suitable only for materials that are not heat sensitive. The heat-sensitive materials, the serum albumin nanoparticles, are produced by emulsifying in cottonseed oil at 25 °C. Then the particles are suspended in cross-linking agents like 2, 3 butadiene or formaldehyde. The resultant nanoparticles are centrifuged and dried by lyophilization [97].

4.19 Methods for Production of Nanoparticles

Several methods are available for the production of nanoparticles: (i) electrochemical deposition, (ii) sol–gel process, (iii) gas-phase condensation, (iv) plasma enhanced vapour decomposition, and (v) sputtering technique.

4.20 Electrochemical Deposition

Electrochemical deposition is a metallic coating on a base material by electrochemical reduction of metal ions. It is mainly to reduce friction and wear, achieve desired electrical properties, reduce corrosion, and increase heat resistance. It is sometimes called electroplating and electro-crystallization, due to the crystalline nature of the coating. In electrodeposition, the cathode and the anode are dipped in the electrolyte, mostly metal-containing liquid. When electric current passes through the electrodes, the metal ions are deposited on the metal [98].

4.21 Sol–Gel Method

Wet-chemical technique of production of nanomaterial is also called chemical solution deposition. The precursors, carbonates or acetates or nitrates are dissolved in deionized water. The preparation of colloidal substances using the starting materials is called a gel. Polyvinyl alcohol, a gelling agent, is added to the starting material to produce a gel. A coating is made on a substrate with controlled temperature, pH, and viscosity. Finally, the coated product is annealed at suitable temperatures [99, 100].

4.22 Gas-Phase Condensation Methods

In this method, the particle size of the nanoparticle is smaller than 10 nm. In inert gas environmental conditions, the metals are introduced at lower pressure (0.5–4 Torr) in a temperature-controlled alumina crucible. Due to the metal collision, the metal vapours are cooled rapidly in the presence of inert gas and the super-saturated vapour undergoes homogeneous nucleation. The size of the particle is influenced by the pressure of inert gas and atomic weight [101].

4.23 Plasma-Enhanced Chemical Vapour Deposition

This method is widely adopted for making the thin film production, where in the conventional chemical vapour deposition (CVD), thermally decompose the inert gas at 500–1,000 °C, whereas in the PECVD method, even at room temperature, the inert gases are dissociated in the presence of plasma. For sensitive materials, damaged by temperature, PECVD can be used where plasma can modify the properties of the film due to the concurrent "ion bombardment" of plasma [102]. Direct and remote methods are commonly practiced in PECVD. In the direct method, the substrate– inert gas, and the gas dilutants are directly fed into the reaction chamber, whereas in the remote method, the plasma is generated separately from the region of deposition and then the inert gases are introduced into the plasma chamber [103].

4.24 Sputtering

One of the efficient methods to produce the nanoparticles is sputtering at high pressure using a standard magnetron DC/RF to sputter the metal at high pressure (0.23–1.5 Torr) in an inert gas atmosphere, without using any solvents [104, 105]. At this high pressure, the gas-phase nucleation is formed. The nanoparticles formed on the gold finger are collected using the scraper.

5 Conclusion

Recommendations of eco-friendly, herbal-based finishes are widely accepted by manufacturers and users in a big way, considering their harmless nature and availability. And, many unexplored natural resources exist in different geographical locations that can potentially be used to finish the textile materials and meet the requirements. Needless to say, newer techniques like nanoencapsulation offer higher efficiency and enhances the durability of the finish. Combinations of such newer methods and naturally available finishing agents provide novel solutions that are sustainable and reduce hazardous impacts.

References

1. Wiegand I, Hilpert K, Hancock REW (2008) Agar and broth dilution methods to determine the minimal inhibitory concentration (MIC) of antimicrobial substances. Nat Protoc 3:163–175. https://doi.org/10.1038/nprot.2007.521
2. Cepae BA (2009) WHO monographs on selected medicinal plants
3. Chopra RN, Nayer SL, Chopra IC (1956) Glossary of Indian medicinal plants. CSIR, New Delhi
4. Kirtikar KR, Basu BD (1975) Medicinal plants. In: Blatter E, Cains JF, Mhaskar KS (eds) Medicinal plants. Vivek Vihar, New Delhi, p 536
5. Koul O, Isman MB, Ketkar CM (1990) Properties and uses of neem, *Azadirachta indica*. Can J Bot 68:1–11. https://doi.org/10.1139/b90-001
6. Biswas K, Chattopadhyay I, Banerjee RK, Bandyopadhyay U (2002) Biological activities and medicinal properties of neem (*Azadirachta indica*). Curr Sci 82:1336–1345. http://www.iisc.ernet.in/currsci/jun102002/1336.pdf
7. Siddiqui S (1942) A note on the isolation of three new bitter principles from the nim oil. Curr Sci 11:278–279
8. Thilagavathi G, Kannaian T (2008) Dual antimicrobial and blood repellent finishes for cotton hospital fabrics. Indian J Fibre Text Res 33:23–29
9. Thilagavathi G, Rajendrakumar K, Rajendran R (2005) Development of ecofriendly antimicrobial textile finishes using herbs. Indian J Fibre Text Res 30:431–436
10. Chandrasekaran K, Ramachandran T, Vigneswaran C (2012) Effect of medicinal herb extracts treated garments on selected diseases. Indian J Tradit Knowl 11:493–498
11. Guo C, Zhou L, Lv J (2013) Effects of expandable graphite and modified ammonium polyphosphate on the flame-retardant and mechanical properties of wood flour-polypropylene composites. Polym Polym Compos 21:449–456. https://doi.org/10.1002/app

12. Saraswathi R, Krishnan PN, Dilip C (2010) Antimicrobial activity of cotton and silk fabric with herbal extract by micro encapsulation. Asian Pac J Trop Med 3:128–132. https://doi.org/10.1016/S1995-7645(10)60051-X

13. Hooda S, Yadav N, Sikka VK (2013) Eco-friendly antimicrobial finish for wool fabric 5:11–16. https://doi.org/10.1080/09751270.2013.11885209

14. Thilagavathi G, Bala SK, Kannaian T (2007) Microencapsulation of herbal extracts for microbial resistance in healthcare textiles. Indian J Fibre Text Res 32:351–354

15. Subramani K, Murugan V, Kolathupalayam Shanmugam B, Rangaraj S, Palanisamy M, Venkatachalam R, Suresh V (2017) An ecofriendly route to enhance the antibacterial and textural properties of cotton fabrics using herbal nanoparticles from *Azadirachta indica* (neem). J Alloy Compd 723:698–707. https://doi.org/10.1016/j.jallcom.2017.06.242

16. Rajendran R, Radhai R, Kotresh TM, Csiszar E (2013) Development of antimicrobial cotton fabrics using herb loaded nanoparticles. Carbohyd Polym 91:613–617. https://doi.org/10.1016/j.carbpol.2012.08.064

17. Ahmed H, Rajendran R, Balakumar C (2012) Nanoherbal coating of cotton fabric to enhance antimicrobial durability. Appl Chem 45:7840–7843

18. Anitha S, Vaideki K, Jayakumar S, Rajendran R (2015) Enhancement of antimicrobial efficacy of neem oil vapour treated cotton fabric by plasma pretreatment. Mater Technol 30:368–377. https://doi.org/10.1179/1753555715Y.0000000015

19. Joshi B, Lekhak S, Sharma A (2010) Antibacterial property of different medicinal plants: *Ocimum sanctum, Cinnamomum zeylanicum, Xanthoxylum armatum* and *Origanum majorana*. Kathmandu Univ J Sci Eng Technol 5:143–150. https://doi.org/10.3126/kuset.v5i1.2854

20. Gupta SK, Prakash J, Srivastava S (2002) Validation of traditional claim of Tulsi, *Ocimum sanctum Linn*. As a medicinal plant. Indian J Exp Biol 40:765–773

21. Ranganathan S, Balajee SAM (2000) Anti-cryptococcus activity of combination of extracts of *Cassia alata* and *Ocimum sanctum*. Mycoses 43:299–301

22. Sathianarayanan MP, Bhat NV, Kokate SS, Walunj VE (2010) Antibacterial finish for cotton fabric from herbal products. Indian J Fibre Text Res 35:50–58

23. Murugesh Babu K, Ravindra KB (2015) Bioactive antimicrobial agents for finishing of textiles for health care products. J Text Inst 106:706–717. https://doi.org/10.1080/00405000.2014.936670

24. Ammayappan L, Moses JJ (2009) Study of antimicrobial activity of aloevera, chitosan, and curcumin on cotton, wool, and rabbit hair. Fibres Polym 10:161–166. https://doi.org/10.1007/s12221-009-0161-2

25. Jasso De Rodríguez D, Hernández-Castillo D, Rodríguez-García R, Angulo-Sánchez JL (2005) Antifungal activity in vitro of Aloe vera pulp and liquid fraction against plant pathogenic fungi. Ind Crops Prod 21:81–87. https://doi.org/10.1016/j.indcrop.2004.01.002

26. Arunkumar S, Muthuselvam M (2009) Analysis of phytochemical constituents and antimicrobial activities of *Aloe vera L*. against clinical pathogens. World J Agric Sci 5:572–576

27. Wasif SK, Rubal S (2007) 6th International conference—TEXCSI. In: Liberec, Czech Republic. Liberec, Czech Republic

28. Ali SW, Purwar R, Joshi M, Rajendran S (2014) Antibacterial properties of *Aloe vera* gel-finished cotton fabric. Cellulose 21:2063–2072. https://doi.org/10.1007/s10570-014-0175-9

29. Jothi D (2009) Experimental study on antimicrobial activity of cotton fabric treated with aloe gel extract from *Aloe vera* plant for controlling the Staphylococcus aureus (bacterium). Afr J Microbiol Res 3:228–232

30. Singh RP, Chidambara Murthy KN, Jayaprakasha GK (2002) Studies on the antioxidant activity of pomegranate (*Punica granatum*) peel and seed extracts using in vitro models. J Agric Food Chem 50:81–86. https://doi.org/10.1021/jf010865b

31. Prashanth D, Asha MK, Amit a (2001) Antibacterial activity of *Punica granatum*. Fitoterapia 72:171–173. https://doi.org/10.1016/S0367-326X(00)00270-7

32. Ghaheh FS, Nateri AS, Mortazavi SM, Abedi D, Mokhtari J (2012) The effect of mordant salts on antibacterial activity of wool fabric dyed with pomegranate and walnut shell extracts. Color Technol 128:473–478. https://doi.org/10.1111/j.1478-4408.2012.00402.x

33. Mahesh S, Reddy AHM, GVK (2011) Studies on antimicrobial textile finish using certain plant natural products. In: International conference on advances in biotechnology and pharmaceutical sciences, pp 253–258
34. Kirtikar KR, Basu BD (1944) Indian medicinal plants. BIO-GREEN BOOKS
35. Gobalakrishnan M, Saravanan D (2020) Antimicrobial activity of *Gloriosa superba, Cyperus rotundus* and *Pithecellobium dulce* with different solvents. Fibres TextEs East Eur 28:67–71. https://doi.org/10.5604/01.3001.0014.0937
36. Tiwari OP, Tripathi YB (2007) Antioxidant properties of different fractions of Vitex negundo Linn. Food Chem 100:1170–1176. https://doi.org/10.1016/j.foodchem.2005.10.069
37. Kumar P, Kumaravel S, Lalitha C (2010) Screening of antioxidant activity, total phenolics and GC–MS study of *Vitex negundo*. Afr J Biochem Res 4:191–195
38. Karunamoorthi K, Ramanujam S, Rathinasamy R (2008) Evaluation of leaf extracts of *Vitex negundo L.* (Family: Verbenaceae) against larvae of *Culex tritaeniorhynchus* and repellent activity on adult vector mosquitoes. Parasitol Res 103:545–550. https://doi.org/10.1007/s00436-008-1005-5
39. Sathiamoorthy B, Gupta P, Kumar M, Chaturvedi AK, Shukla PK, Maurya R (2007) New antifungal flavonoid glycoside from *Vitex negundo*. Bioorg Med Chem Lett 17:239–242. https://doi.org/10.1016/j.bmcl.2006.09.051
40. Alam MI, Gomes A (2003) Snake venom neutralization by Indian medicinal plants (*Vitex negundo* and *Emblica officinalis*) root extracts. J Ethnopharmacol 86:75–80. https://doi.org/10.1016/S0378-8741(03)00049-7
41. Mohanraj S, Vanathi P, Sowbarniga N, Saravanan D (2012) Antimicrobial effectiveness of *Vitex negundo* leaf extracts. Indian J Fibre Text Res 37:389–392
42. Mukherjee PK, Kumar V, Mal M, Houghton PJ (2007) Acetylcholinesterase inhibitors from plants. Phytomedicine 14:289–300. https://doi.org/10.1016/j.phymed.2007.02.002
43. Ponnusamy S, Gnanaraj WE, Antonisamy JM, Selvakumar V, Nelson J (2010) The effect of leaves extracts of *Clitoria ternatea Linn* against the fish pathogens. Asian Pac J Trop Med 3:723–726. https://doi.org/10.1016/S1995-7645(10)60173-3
44. Anand SP, Doss A, Nandagopalan V (2011) Antibacterial studies on leaves of *Clitoria Ternatea Linn.*—A high potential medicinal plant. Int J Appl Biol Pharm Technol 2:453–456
45. Malabadi RB, Naik SL, Meti NT, Mulgund GS (2012) Silver nanoparticles synthesized by in vitro derived plants and callus cultures of Clitoria ternatea; evaluation of antimicrobial activity. Res Biotechnol 3:26–38
46. Malabadi RB, Mulgund GS, Meti NT, Nataraja K (2012) Antibacterial activity of silver nanoparticles synthesized by using whole plant extracts of *Clitoria ternatea*. Res Pharm 2:10–21
47. Morton JF (1992) Country Borage (*Coleus amboinicus Lour.*). J Herbs Spices & Med Plants 1:77–90. https://doi.org/10.1300/J044v01n01_09
48. Mathew S, Abraham TE (2006) In vitro antioxidant activity and scavenging effects of *Cinnamomum verum* leaf extract assayed by different methodologies. Food Chem Toxicol 44:198–206. https://doi.org/10.1016/j.fct.2005.06.013
49. Gobalakrishnan M, Saravanan D (2017) Antimicrobial activity of *Coleus ambonicus* herbal finish on cotton fabric. Fibres TextEs East Eur 25:106–109. https://doi.org/10.5604/01.3001.0010.2854
50. Khanum H, Ramalakshmi K, Srinivas P, Borse BB (2011) Synergistic antioxidant action of oregano, ajowan and borage extracts. Food Nutr Sci 02:387–392. https://doi.org/10.4236/fns.2011.25054
51. Chandrappa MS, Harsha R, Dinesha R, Gowda TSS (2010) Antibacterial activity of *Coleus aromaticus* leaves. Int J Pharm Pharm Sci 2:63–66
52. Gurgel APAD, da Silva JG, Grangeiro ARS, Xavier HS, Oliveira RAG, Pereira MSV, de Souza IA (2009) Antibacterial effects of *Plectranthus amboinicus* (*Lour.*) spreng (*Lamiaceae*) in methicillin resistant *Staphylococcus aureus* (MRSA). Lat Am J Pharm 28:460–464. https://doi.org/10.1.1.500.9302

53. Narayanan KB, Sakthivel N (2010) Phytosynthesis of gold nanoparticles using leaf extract of *Coleus amboinicus Lour*. Mater Charact 61:1232–1238. https://doi.org/10.1016/j.matchar.2010.08.003

54. Jayaweera DMA (1982) Medicinal plants used in Ceylon, vol 3. National Science Council of SriLanka, Colombo

55. Singh AK (2006) Flower crops: cultivation and management. New India Publishing Agency, New Delhi

56. Rajak RC, Rai MK (1990) Herbal medicines, biodiversity and conservation strategies. International Book Distributors

57. Raut NA, Gaikwad NJ (2006) Antidiabetic activity of hydro-ethanolic extract of *Cyperus rotundus* in alloxan induced diabetes in rats. Fitoterapia 77:585–588. https://doi.org/10.1016/j.fitote.2006.09.006

58. Parekh J, Chanda S (2006) In-vitro Antimicrobial Activities of extracts of *Launaea procumbens Roxb*. (Labiateae), *Vitis vinifera L.* (Vitaceae) and *Cyperus rotundus L.* (Cyperaceae). Afr J Biomed Res 9:89–93. https://doi.org/10.4314/ajbr.v9i2.48780

59. Shanmugakumaran SD, Amerjothy S, Balakrishna K, Kumar MSV (2005) Antimycobacteri properties of leaf extracts of *Pithecellobium dulce* Benth. Indian Drugs 42:392–395

60. Khatoon S, Singh S, Singh M, Kumar V, Rawat AS, Mehrotra S (2010) Antimicrobial screening of ethnobotanically important stem bark of medicinal plants. Pharmacogn Res 2:254–258. https://doi.org/10.4103/0974-8490.69127

61. Mukesh Kumar KN, JSD (2013) Phytochemical analysis and antimicrobial efficacy of leaf extracts of *Pithecellobium dulce*. Asian J Pharm Clin Res 6:70–76

62. Ramachandran J (2008) Herbs of siddha medicine/the first 3D book on herbs. Murugan Patthipagam, Chennai, India

63. Burkill HM (1985) The useful plants of west tropical africa, 2nd edn. Royal Botanic Gardens, Kew, UK

64. Nahrstedt A, Hungeling M, Petereit F (2006) Flavonoids from *Acalypha indica*. Fitoterapia 77:484–486. https://doi.org/10.1016/j.fitote.2006.04.007

65. Govindarajan M, Jebanesan A, Reetha D, Amsath R, Pushpanathan T, Samidurai K (2008) Antibacterial activity of *Acalypha indica L.* Eur Rev Med Pharmacol Sci 12:299–302. https://doi.org/10.17485/ijst/2008/v1i6/29582

66. Balarabe S, Habibu S, Muhammad S, Ladan M, Haruna A, Baba I, Nafiu S (2017) Dyeing and antibacterial finishing of cotton fabric using diospyros mespiliformis leaves extracts 13:174–177

67. Benoit JP, Marchais H, Rolland H, Van de Velde V (1996) Biodegradable microspheres: advances in production technology. In: SB (ed) Microencapsulation: methods and industrial applications. Marcel Dekker, Inc., pp 35–72

68. Arshady R (1991) Preparation of biodegradable microspheres and microcapsules: 2. Polyactides and related polyesters. J Control Release 17:1–21. https://doi.org/10.1016/0168-3659(91)90126-X

69. Kent JS, Sanders LM, Lewis DH, Tice TR (1982) Microencapsulation of water-soluble polypeptides. Eur Patent 052:510

70. Ruiz J, Busnel J, Benoît J (1990) Influence of average molecular weights of poly(DL-lactic acid-CO-glycolic acid) copolymers 50/50 on phase separation and in vitro drug release from microspheres. Pharm Res 07:928–934. https://doi.org/10.1023/A:1015945806917

71. Bodmeier R, Chen H (1988) Preparation of biodegradable poly($\pm$)lactide microparticles using a spray-drying technique. J Pharm Pharmacol 40:754–757. https://doi.org/10.1111/j.2042-7158.1988.tb05166.x

72. Baras B, Benoit M-A, Gillard J (2000) Parameters influencing the antigen release from spray-dried poly(DL-lactide) microparticles. Int J Pharm 200:133–145. https://doi.org/10.1016/S0378-5173(00)00363-X

73. Prior S, Gamazo C, Irache J, Merkle H, Gander B (2000) Gentamicin encapsulation in PLA/PLGA microspheres in view of treating Brucella infections. Int J Pharm 196:115–125. https://doi.org/10.1016/S0378-5173(99)00448-2

74. Wagenaar BW, Müller BW (1994) Piroxicam release from spray-dried biodegradable microspheres. Biomaterials 15:49–54. https://doi.org/10.1016/0142-9612(94)90196-1

75. Blanco-Príeto MJ, Besseghir K, Zerbe O, Andris D, Orsolini P, Heimgartner F, Merkle HP, Gander B (2000) In vitro and in vivo evaluation of a somatostatin analogue released from PLGA microspheres. J Control Release 67:19–28. https://doi.org/10.1016/S0168-365 9(99)00289-8

76. Debenedetti PG, Tom JW, Sang-Do Y, Gio-Bin L (1993) Application of supercritical fluids for the production of sustained delivery devices. J Control Release 24:27–44. https://doi.org/10.1016/0168-3659(93)90166-3

77. Thiering R, Dehghani F, Foster NR (2001) Current issues relating to anti-solvent micronisation techniques and their extension to industrial scales. J Supercrit Fluids 21:159–177. https://doi.org/10.1016/S0896-8446(01)00090-0

78. Jung J, Perrut M (2001) Particle design using supercritical fluids: literature and patent survey. J Supercrit Fluids 20:179–219. https://doi.org/10.1016/S0896-8446(01)00064-X

79. Boswell G, Scribner R (1973) Polylactide-drug mixtures, pp 1–8

80. Couvreur P, Dubernet C, Puisieux F (1995) Controlled drug delivery with nanoparticles: current possibilities and future trends. Eur J Pharm Biopharm 41:2–13

81. Kreuter J (1990) Large-scale production problems and manufacturing of nanoparticles. In: Tyle P (ed) Specialized drug delivery system. Marcel Dekker, New York, pp 257–266

82. Ekman B, Sjöholm I (1978) Improved stability of proteins immobilized in microparticles prepared by a modified emulsion polymerization technique. J Pharm Sci 67:693–696. https://doi.org/10.1002/jps.2600670533

83. Lowe PJ, Temple CS (1994) Calcitonin and insulin in isobutylcyanoacrylate nanocapsules: protection against proteases and effect on intestinal absorption in rats. J Pharm Pharmacol 46:547–552. https://doi.org/10.1111/j.2042-7158.1994.tb03854.x

84. Kreuter J (1991) Nanoparticle-based drug delivery systems. J Control Release 16:169–176

85. El-Samaligy MS, Rohdewald P, Mahmoud HA (1986) Polyalkyl cyanoacrylate nanocapsules. J Pharm Pharmacol 38:216–218. https://doi.org/10.1111/j.2042-7158.1986.tb04547.x

86. Vauthier C, Dubernet C, Fattal E, Pinto-Alphandary H, Couvreur P (2003) Poly(alkylcyanoacrylates) as biodegradable materials for biomedical applications. Adv Drug Deliv Rev 55:519–548. https://doi.org/10.1016/S0169-409X(03)00041-3

87. Couvreur P, Barratt G, Fattal E, Vauthier C (2002) Nanocapsule technologys: a review. Crit Rev Ther Drug Carrier Syst 19:99–134. https://doi.org/10.1615/CritRevTherDrugCarrierSyst.v19.i2.10

88. Al Khouri Fallouh N, Roblot-Treupel L, Fessi H, Devissaguet JP, Puisieux F (1986) Development of a new process for the manufacture of polyisobutylcyanoacrylate nanocapsules. Int J Pharm 28:125–132. https://doi.org/10.1016/0378-5173(86)90236-X

89. Watnasirichaikul S, Davies NM, Rades T, Tucker IG (2000) Preparation of biodegradable insulin nanocapsules from biocompatible microemulsions. Pharm Res 17:684–689. https://doi.org/10.1023/A:1007574030674

90. Montasser I, Fessi H, Coleman A (2002) Atomic force microscopy imaging of novel type of polymeric colloidal nanostructures. Eur J Pharm Biopharm 54:281–284. https://doi.org/10.1016/S0939-6411(02)00087-5

91. Bouchemal K, Briançon S, Perrier E, Fessi H, Bonnet I, Zydowicz N (2004) Synthesis and characterization of polyurethane and poly(ether urethane) nanocapsules using a new technique of interfacial polycondensation combined to spontaneous emulsification. Int J Pharm 269:89–100. https://doi.org/10.1016/j.ijpharm.2003.09.025

92. Tice TR, Gilley RM (1985) Preparation of injectable controlled-release microcapsules by a solvent-evaporation process. J Control Release 2:343–352. https://doi.org/10.1016/0168-365 9(85)90056-2

93. Fessi H, Puisieux F, Devissaguet JP, Ammoury N, Benita S (1989) Nanocapsule formation by interfacial polymer deposition following solvent displacement. Int J Pharm 55:R1–R4. https://doi.org/10.1016/0378-5173(89)90281-0

94. Quintanar-Guerrero D, Allémann E, Fessi H, Doelker E (1998) Preparation techniques and mechanisms of formation of biodegradable nanoparticles from preformed polymers. Drug Dev Ind Pharm 24:1113–1128. https://doi.org/10.3109/03639049809108571

95. Lambert G, Fattal E, Couvreur P (2001) Nanoparticulate system for the delivery of antisense oligonucleotides. Adv Drug Deliv Rev 47:99–112

96. Patil GV (2003) Biopolymer albumin for diagnosis and in drug delivery. Drug Dev Res 58:219–247. https://doi.org/10.1002/ddr.10157

97. Widder KJ, Flouret G, Senye A (1979) Magnetic microspheres: synthesis of a novel parental drug carrier. J Pharm Sci 68:79–82

98. Shinomiya T, Gupta V, Miura N (2006) Effects of electrochemical-deposition method and microstructure on the capacitive characteristics of nano-sized manganese oxide. Electrochim Acta 51:4412–4419. https://doi.org/10.1016/j.electacta.2005.12.025

99. Callister WDJ, Rethwisch DG (2004) Mater Sci 22:0–990

100. Lu C-H, Jagannathan R (2002) Cerium-ion-doped yttrium aluminum garnet nanophosphors prepared through sol-gel pyrolysis for luminescent lighting. Appl Phys Lett 80:3608–3610. https://doi.org/10.1063/1.1475772

101. Eilers H, Tissue BM (1995) Synthesis of nanophase ZnO, Eu_2O_3, and ZrO_2 by gas-phase condensation with cw-CO_2 laser heating. Mater Lett 24:261–265. https://doi.org/10.1016/0167-577X(95)00112-3

102. Malesevic A, Vitchev R, Schouteden K, Volodin A, Zhang L, Tendeloo G Van, Vanhulsel A, Haesendonck C Van (2008) Synthesis of few-layer graphene via microwave plasma-enhanced chemical vapour deposition. Nanotechnol 19:305604. https://doi.org/10.1088/0957-4484/19/30/305604

103. Teo KBK, Lee S-B, Chhowalla M, Semet V, Binh VT, Groening O, Castignolles M, Loiseau A, Pirio G, Legagneux P, Pribat D, Hasko DG, Ahmed H, Amaratunga GAJ, Milne WI (2003) Plasma enhanced chemical vapour deposition carbon nanotubes/nanofibres how uniform do they grow? Nanotechnology 14:204–211. https://doi.org/10.1088/0957-4484/14/2/321

104. Seol J, Lee S, Lee J, Nam H, Kim K (2003) Electrical and optical properties of CuZnSnS thin films prepared by RF magnetron sputtering process. Sol Energy Mater Sol Cells 75:155–162. https://doi.org/10.1016/S0927-0248(02)00127-7

105. Heo CH, Lee S-B, Boo J-H (2005) Deposition of TiO_2 thin films using RF magnetron sputtering method and study of their surface characteristics. Thin Solid Films 475:183–188. https://doi.org/10.1016/j.tsf.2004.08.033

An Overview of Preparation, Processes for Sustainable Denim Manufacturing

M. R. Srikrishnan and S. Jyoshitaa

Abstract Fashion changes, but denim is here to stay. It is a popular field of fashion despite the periodical changes. Currently, denim is not just a garment, but a fabric for an entire lifestyle. The sad fact is that denim is one of the biggest contributors to pollution from the fashion industry. Currently, the focus is on environmental friendly fashion. Even so, denim industries are still a fast-growing part of the fashion market. The difficulties that lie beneath the vast denim industry range from massive amounts of secondhand trash, unsaleable stock and denim waste that must be processed again. Repurposing denim waste into attractive things, on the other hand, is only a minor portion of the sustainable process. The future of denim is also promising to owe to innovations. Denim continues to be revived throughout the fashion cycle in order to generate new fashion trends. Today's buzzwords are sustainability and recycling, and everyone involved in the apparel supply chain, from manufacturers to customers, is striving to make this a reality. New technologies are being developed to recycle denim fibres while maintaining their quality. Government support has also expanded to include many solid waste management programmes, landfill reduction through reuse and recycling, and environmental laws and regulations. There are many options available to denim enthusiasts today such as Advanced and organic denim, less polluting fabric dyeing and washing processes, zero water technologies, oxygen and ozone washes, recycling processes, eco-denim projects. This chapter focuses on the different sustainability aspects that can be followed in each and every phase of denim product lifecycle, i.e. from raw material to disposal phase and also focuses on the recent studies in the area of sustainable denim development.

Keywords Denim · Sustainability · Phases of denim lifecycle · Denim technologies · Denim washing · Denim care · Upcycling · Repurposing of denim

M. R. Srikrishnan (✉) · S. Jyoshitaa
Department of Fashion Technology, PSG College of Technology, Peelamedu, Coimbatore 641004, India
e-mail: mrs.fashion@psgtech.ac.in

© The Author(s), under exclusive license to Springer Nature Singapore Pte Ltd. 2022   119
S. S. Muthu (ed.), *Sustainable Approaches in Textiles and Fashion*,
Sustainable Textiles: Production, Processing, Manufacturing & Chemistry,
https://doi.org/10.1007/978-981-19-0538-4_5

1 Introduction

Denim is a distinctive fabric made from indigo-dyed cotton yarn in the warp and undyed cotton yarn in the weft, woven in a warp-faced twill pattern. This makes the front of the fabric blue and the back white. Denim is a fabric which is suitable for all seasons. It is a very fashionable fabric. It has undergone constant evolution to remain relevant in a rapidly changing industry. Denim is becoming more of a whole lifestyle fabric than just a clothing. In the world of fashion, denim jeans have sparked a blue revolution. Consumers have an inexhaustible fondness for denim, making it a popular trend. The textile sector heavily relies on denim. It is a timeless classic that will almost likely never go out of style. Denim can be used to create a variety of clothing and styles, including traditional and vintage chinos, skinny shorts, capris, skirts and dresses. It is also available in a variety of specialty finishes and treatments, such as printed denim, water-repellent denim and recycled denim.

Because of the large amounts of mineral substances used to process denim, the denim processing industry continues to deliver certain environmental pollution. So, each stage of the denim manufacturing process has an environmental impact, such as energy and water use, CO_2 emissions, and waste. Therefore, pollution by the denim industry includes air pollution, water pollution, noise pollution, etc. depending on the production stage.

Solid waste is an integral part of textile production, and denim waste consists of pre-consumption waste (clean waste) and post-consumption waste (used waste). Metal accessories such as rivets and zippers used in jeans need to be properly recycled. At the end of a product's life cycle, the stages of use and disposal depend on consumer behaviour. However, these are important phases in the product life cycle. Consumers also need to be aware that the financial costs associated with making jeans are not the only ones. They are associated with huge environmental costs.

2 Phases of a Denim Life Cycle

The product life cycle is an assessment of a product's whole life cycle, which begins with the extraction of raw materials and concludes with its disposal stage. The major goal of this evaluation is to assess the environmental effects of denim and to evaluate the phases and impact of a product's complete life cycle, from raw material extraction (cradle) through waste treatment (grave), in order to improve its performance throughout its life cycle. Figure 1 is an illustration of all phases in a denim's life cycle.

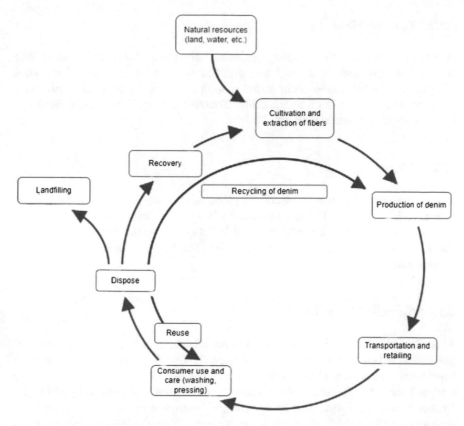

Fig. 1 Phases of a product's life cycle (for denim)

2.1 Raw Material Phase

Functional fibres that meet eco-standards are indispensable for the denim product line. Popular synthetic fibres such as polypropylene, acrylic, polyester and others are gradually being replaced by biodegradable and environmentally friendly fibres.

2.1.1 Organic Cotton

Organic cotton can be cultivated in an ecofriendly manner without using synthetic materials. Cotton farming accounts for 16% of total consumed insecticides. There is not much difference in the cotton fibre length and strength of conventional and organic cotton. Denim manufacturers are moving towards using organic cotton.

2.1.2 Bamboo Fibre

Bamboo is regarded as the ultimate green material because it is available in an endless supply. Bamboo fibres have a high tenacity, whiteness index equivalent to bleached viscose, good antibacterial characteristics, high moisture regain value, high fabric handling and high level of breathability. Therefore, it could be an excellent raw material for denim manufacturing.

2.1.3 Linen

The fashion trend toward an unstructured informal style has reintroduced linen to the spotlight. It's stylish and long-lasting, and it's ideal for fuller skirts, dresses, pants and jackets. The 70/30 cotton/linen blend is both economical and has a great hand and drape. Other bast fibres, such as hemp, jute and their mixtures, are now used to make denim.

2.1.4 Tencel™/Cotton Blend

Lyocell is a cellulosic fibre that outperforms all others in terms of characteristics and looks. It is also made in an environmentally friendly manner. The most well-known and extensively utilised Lyocell fibres are Tencel™ and its non-fibrillating variant Tencel A 100. Tencel™ is a luxurious fabric that is smooth, silky and shiny. Tencel™ retains 85 percent of its dry strength even when wet, making it the only man-made cellulosic fibre that outperforms cotton. Tencel™ denim, which contains at least 40% lyocell, has virtually altered the concept of denim all over the world. The new regenerated fibre is highly durable and extremely comfortable compared to cotton denims. Tencel™ is highly breathable, absorbs moisture and performs better in humid situations than when compared to cotton. It's also less bulky and easy to pack.

2.1.5 Polylactic Acid (PLA) Fibres

Polylactic acid is a naturally existing, biodegradable organic compound found in the bodies of animals, plants and microorganisms. It is not found as such in nature, and therefore it is made in a lab using lactic acid polymerization. Lactic acid is made from corn grains that have been genetically modified. Melt spinning is the process that is used to make polylactic acid fibres. PLA's fundamental polymer chemistry permits it to have superior mechanical characteristics, low flammability, high UV protection index, low refraction index and low moisture regain value.

2.1.6 Soybean Fibres

Soy protein fibre is a liquefied vegetable protein fibre that is extracted from soybeans after oil is extracted. Using sophisticated bioengineering technologies, it is mechanically processed to generate fibres. The fibres are manufactured by wet spinning, stabilised by acetylation, and finally cut into short staples after crimping and thermoforming. A soybean protein fibre combines the advantages of natural fibres with the physical features of synthetic fibres. They are also less expensive and have a high fabric handle, moderate moisture regain, good moisture transmission characteristics, moderate tensile strength, and outstanding anti-crease, easy wash, and quick-drying properties.

2.1.7 Ramie/Cotton Denim

Ramie/cotton blends are available in a variety of combinations with a wide range of prices. Ramie is a plant fibre that is commonly used to prevent wrinkling and give the fabric a silky lustre. Despite the fact that ramie is not as strong as cotton, it works well as a raw material for denim manufacturing when it is blended with other fibres.

2.1.8 Recycled Polyethylene Terephthalate (rPET)

Recycled polyester is made by melting existing plastic and spinning it again to make new polyester fibres. While rPET manufactured from consumer-discarded plastic bottles and containers receives a lot of attention, polyethylene terephthalate can be recycled from both post-industrial and post-consumer sources. Eight to ten used bottles and scrap cotton obtained from manufacturing plant floors are incorporated in each pair of these jeans, which would otherwise be discarded in landfills. This displays the organization's commitment to environmental preservation while also operating profitably. While it sounds simple, there has been a considerable amount of groundbreaking work to assemble jeans from used water bottles and popularise plastic. While recycling plastic seems like an undeniably good idea, the rPET celebration is far from unanimous in the sustainable fashion community.

2.2 Manufacturing Phase

The process of spinning fibres into yarn, yarn into fabric and fabric into clothing accounts for about 70% of the total energy used to make a pair of jeans. Manufacturing jeans is a complicated process that necessitates the use of specialised equipment and technology, as well as complex abilities combined with a high level of creativity and specialised knowledge. Denim's signature indigo colour and faded look are the outcomes of chemically intense and high-water-use treatment processes that can be hazardous to workers' health and the environment. Dedicated designers, consumers, and global interest from denim brands and manufacturers make the process meaningful and profitable. Renowned denim brands are actively promoting techniques such as enzyme finishing, laser etching and ozone treatments that limit the chemical and water intensity of wet processing.

2.2.1 Eco Friendly Dyeing Process

Due to confined manufacturing of herbal indigo, artificial indigo is used. It is not viable to use 100% of the dye, left over dye is discharged into water which affects the ecosystem. Moreover, indigo is insoluble in water, and it calls for a massive amount of alkali (NaOH) and a decreasing agent which includes sodium dithionite ($Na_2S_2O_4$) to make it water soluble. Utilisation of alkali and a decreasing agent may also grow the TDS, BOD. In a few cases, to reap dark colours, sulphur dyes may be used. Sulphides are very risky to dwelling organisms and their health.

To create a sustainable environment, usage of natural dyes should be promoted. Different shades can be achieved on denim using several other dyes and ecofriendly mordants. Onion extract can be used as natural and synthetic mordants.

2.2.2 Environmentally Friendly Bleaching Process

It is far crucial to make use of this opportunity and incorporate methods that doesn't destroy the environmental systems. These alternatives can potentially replace sodium hypochlorite which is the widely used bleaching agent, but none of them can achieve the same effect. Laccases have recently been reported to be used in the bleaching of indigo fabrics. Laccases are a form of oxidoreductase enzyme. In most cases, they don't act alone, a chemical-based mediator is required between the enzyme and the indigo molecules. The indigo is subsequently attacked by free radicals, which turns it into oxidised compounds. Despite being a green method, it cannot compete with traditional hypochlorite bleaching in terms of cost, time or efficiency.

2.2.3 Enzyme Fading

Considering environmental factors and efficiency, environmentally friendly enzymes have been introduced in garment processing. Enzymes are broadly used in denim processing due to their unique properties. Blue denim was faded in the traditional stone washing process by the abrasive action of pumice stones on the fabric, producing a worn-out appearance. Cellulase catalyses the hydrolysis of cellulose molecules' 1,4-beta glycoside bonds. This approach produces less pollution and is easier to treat in effluent plants.

2.2.4 Ozone Fading

Ozone is commonly used as a moderate bleaching and sterilisation agent. However, utilising ozone gas, this method can also be done in a closed room. This approach has a number of advantages such as minimal strength loss, it is a straightforward approach, it is water- and chemical-free, making it environmentally benign, processing requires little energy, and treatment time is limited. After laundry, UV radiation can quickly deozonize the ozonized water. Plasma equipment could also be used to achieve ozone fading. High-energy electrons are developed as a result of plasma treatment. The ·OH radical is also produced by Ultraviolet light. Hydroxyl radical can oxidise indigo dye molecules (RH), releasing extremely reactive organic radicals R· that can be further oxidised. As a result, the indigo coloured textile's colour fading effect is obtained. As a result, no noticeable shade change occurs.

2.2.5 Laser Fading

sThe laser works by generating a large amount of heat. The material is subjected to intense heating in a very small area within the targeted region. Because of the liquid's surface tension, some of the molten liquid seeks to move. The remaining liquid boils and releases fumes as it heats up quickly. Another phase transition occurs, this time from liquid to gas. For several years, the CO_2 laser treatment has been used in various areas of the textile industry since it allows for quick surface pattern design with excellent precision, highly desired effects, and a variety of sizes and intensities without causing significant damage to the bulk properties of the textile materials. Laser Marking (just the surface of the fabric is treated, fading), Laser fading is the popular dry technique used for denim processing nowadays using these lasers. It has been widely utilised to replace several traditional dry techniques like sandblasting, hand sanding, destroying and grinding, all of which have the potential to be hazardous and deleterious in some way. Apart from that, laser systems are employed in fashion design, pleating, cutting and fabric surface modification to impart a unique finish. Laser fading has superior precision and efficiency. When we compare laser fading to hand dry procedures, we get some points as follows:.

- The manual system produces a finer effect than the laser system.
- The manual system has a finer hand feel than the laser system.
- Hairiness is less in the Laser system when compared to manual system.
- Laser system production costs are higher.
- Laser system requires small working area.
- The Laser system has a high level of design consistency.
- Laser faded denim has higher tearing strength (May differ in case of stretch denim).
- The Laser system has a 0% rejection rate, whereas the Manual system has a 5% rejection rate.
- Required manpower—Laser: Manual = 1:3 (Approximately).
- Laser systems have a higher power requirement (approximately 7.5 kWh).
- If the laser beam came into touch with the skin or eyes, it would be dangerous.

2.2.6 Water Jet Fading Technology

Water jet treatment is used in denim garments to create patterns as well as to enhance its surface finish, texture and durability. In this process, either one or both the surfaces of the denim are treated with hydro-jet nozzles. The fluid impact energy and the dye used in the fabric determine the extent of the dye washout, clarity of the pattern and the surface finish. Especially with denims dyed with blue indigo dye, better results are observed. This denim processing is economic and pollution free when a simple water recycling process is carried out. Colour washout in the exposed areas gives a faded effect without any blur or loss in strength and durability of the fabric. Even excessive warp shrinkage is avoided.

2.2.7 Denim Washing Using Natural Ingredients

Garment washing is done in denim to achieve effects such as colour fading with or without patchiness, crinkles, seam puckering, hairiness, De piling, softened-hand feel, stabilised dimensions and so on. Different types of garment wash include bleach wash, enzyme wash, stone wash, acid wash, etc. As these are chemical based, they create hazard to the environment. So, it's high time to bring in natural ingredients for denim washing. Because of its detergency power, soapnut (natural alternative) was used as a de-sizing agent. The sun has the ability to bleach a variety of materials. Under sunlight, tamarind and lemon have been used as bleaching agents. In all of the tests, the denim fabrics washed with lemon and tamarind performed better than the sample washed with calcium hypochlorite and were nearly identical to the sample washed with enzyme. As an outcome, it can be noted that lemon and tamarind have valuable potential as natural reagents in denim wash.

2.2.8 Denim Washing Using Rubber Shoe Sole

Although washing serves as an essential finish to the denim industry, as enhancing fading effects brings softness and comfort, all these exclusively pollute the environment from the fashion market. For instance, stone washing causes fabric degradation, equipment wear, and tear, grit deposition in the effluent plant, increased labour costs to separate the pumice stone granules from the pockets, and it also damages the expensive washing machines, etc. On the other hand, used rubber shoe soles are tossed away or burned, contributing to ozone layer depletion. In this cornerstone, a post-used rubber shoe sole could be used to minimise the negative impact on nature, as the shoe sole provides fabric characteristics similar to a stonewash. One of the fundamental advantages of using rubber shoe sole is that it creates less impact on nature. This alternative method has given best results without compromising on the trendy looks. The work area has also experienced lesser decibels, which is (68–72 db) less compared to stonewash which is (89–90 db). Hence, post-used shoe soles are comparatively effective for denim washing and for creating sustainable environment.

2.3   Consumer Use Phase

Surprisingly, a huge chunk of the environmental impact of a pair of jeans falls on the consumer. The way you launder and care for your jeans, while also increasing its lifetime minimises the denim's ecological footprint. We tend to overwash the denim by force of habit. Participants of a 2012 study wore the same pair of unwashed jeans for three months without any ill effects. Airing and spot cleaning are more than sufficient to remove stains and smells. Each person can develop their own patina through a lifetime of wear. According to experts, jeans should not be machine washed. Follow cold wash cycles and line drying to improve the life of your jeans. President Chip Bergh himself admitted that he never puts his jeans in the wash and spot cleans or hand washes his jeans himself. The gigantic amount of water in the washing machine may cause the denim to fade. Regular washing may help to shrink the jeans back down after being stretched due to wear.

Denim brands are looking forward to embracing longevity and durability. Nudie offers repair services for jeans while Levi Strauss promotes a personal connection to one's clothing.

2.3.1   How to Take Care of Your Denim?

- **Spot clean**

  Spot clean the denim with a toothbrush and a mixture of water and a gentle detergent for small food stains. This keeps the dye intact and also reduces water consumption.
- **Freeze**

  To kill bacteria, fold your jeans, put them in an airtight freezer bag, and place them in the refrigerator freezer overnight.
- **Try a neutralizer spray**

  Dry wash sprays can freshen up worn jeans. Aerosol sprays and even denim-specific sprays protect the denim fibres from wear and tear and slow down fading.
- **Hand wash**

  To clean thoroughly, jeans can be hand washed, especially with cold water.
- **Wash with cold water**

  For a machine wash, the jeans can be turned inside out and washed with similar colours on a cold setting. The zipper should also be unzipped. This protects the metal parts and reduces the effect of fading.

2.3.2 Upcycling of Denims

When products are dismantled or broken down, their value is lost, and the materials may lose quality in some way as a result of the process. This is commonly referred to as recycling–downcycling. This is a regenerative process that produces cleaner end products that may be mixed with other materials to create innovative products. Polyester fibres made from PET bottles are a good example of down cycling, although the materials produced have structural flaws such as strength loss and a lower melting point than virgin polyester fibres. However, technological advancements have aided the plastics industry, with processes such as the conversion of waste plastics into paramagnetic conducting microspheres, carbon nanoparticles by high temperature and chemical vapour deposition, and treating polymers with 150 kGy electron beam dose levels to increase bending strength and elasticity. Aesthetic and technological prowess can be used to create a wide range of items. Handwoven cotton and denim rugs, table runners, accent rugs and door mats can all be woven into machine-washable items. A never-ending list of goods that have been upcycled from denim jean includes hand-quilted blankets, pillow shams, cushion covers, art objects, wall hangers, pot holders, tote bags, costume jewellery, apparels and accessories.

2.4 Repurposing of Denims

Deconstruction and Reconstruction is a type of upcycling process in which new clothes are made from waste formed by previously worn garments. This method entails deconstructing garments first and then reconstructing the waste materials into new designs. We consume 400% more clothing today than we did two decades ago. Waste increases in tandem with rising fashion consumption. Many of the clothes that end up in landfills are still in good condition and contain valuable materials. Designers can reconstruct old and unwanted clothing to create something new and desirable through the process of reconstruction. The environmental advantages of reconstruction are enormous. When you take into account the Earth's limited supply of natural resources, we simply cannot continue at our current rates of consumption because resources will run out at some stage in the future. As a result, at some point, recycling and reconstruction will no longer be optional; they will be required.

The well-known concept for the closed-loop mindset is reuse, repair, recycle, redesign and reimagine. The most essential need is to increase the use phase of a garment. Glue jeans offer seams that are glued instead of being stitched in the seam. This facilitates mechanical recycling without much hindrances.

2.5   Post-consumer Waste

End users purchase significant quantities of denim jeans. A total of 3.6 billion pairs of jeans are produced each year. The overall textile consumption of jeans is 2.16 million metric tons per year, assuming an average pair of jeans weighing 600 g. After one or two years of use, this quantity will be discarded. This means that there is a theoretical annual capacity of 2.16 million tons of post-consumer denim trash. Only a small portion of this waste gets collected and recycled. It is estimated that only 35–50% of textiles are collected in Western Europe.

The collecting of post-consumer jeans is a major issue. Despite the existence of collecting facilities (curbside collection, textile waste containers) in many countries, many consumers throw their jeans as solid municipal waste. This garbage is disposed of in landfills or burnt. The jeans get moist and nasty when mixed with other debris, and high-end recycling is no longer an option.

Collected jeans are generally sold to textile sorting businesses. They manually sift rewearable jeans (particularly branded jeans) for sale in thrift stores and Third World countries. The majority of non-rewearable jeans are sold to textile recycling businesses. More effort is being made in several countries to increase the amount of textile material (particularly jeans) collected. National and local authorities are working together with charities and commercial textile waste collectors to convince consumers to separate their textiles and donate them to charities and clothing banks.

2.5.1   Recycling Technique

There are many ways to recycle waste denim. However, in practice, only a few are encountered—these are the economically and ecologically preferred methods. Recycling of denim is primarily done by breaking down post-consumer materials into fibres that can be reused in inferior products. The classified denim fractions are sold to textile shredders and this denim fraction is shredded into fibres without prior treatment. The fibres are processed into non-woven fabrics for special applications. The automotive industry is one of the main users of shredder denim non-woven fabrics for sound and heat insulation. Nonwovens for home insulation, as a substitute for mineral wool, are also made from these fibres.

High-quality recycling of post-consumer denim waste may require additional steps to obtain good-quality, pure fibres that can be spun into high-quality yarns. The main problems of post-consumer denim in high-end recycling are metal buttons, rivets, labels which resemble leather, labels used for composition and care instructions, thick seams. Non-textile parts must be disposed of as recycled fibres must not contain impurities such as metals. Parts that are difficult to untangle are fed back into the process until they finally turn into fibres. Buttons, zippers and most rivets can be removed by trimming and using only the trouser legs; the upper part, with its buttons, zippers, rivets and labels, is either unused or used in applications of inferior quality products. The legs of the jeans, which account for about 50% of the total weight, are ripped and untangled. The resulting fibres can be carded to obtain virtually contaminant-free fibres, yarn ends and fabric scraps. This material is generally suitable for spinning, although it may be necessary to add virgin fibres to obtain a yarn with adequate mechanical properties. If the entire post-consumer jeans are to be used in the shredding and unravelling process, the shredding machine must be capable of distinguishing non-textile portions from textile fibres. Step cleaners are used to separate heavy components from fibres in most shredding machines of this sort, such as the Laroche Jumbo.

Metal detectors are also used to remove metal buttons and zippers from apparel. Most buttons and other metal bits can be removed in this manner, but there is no guarantee that all non-textile elements will be removed. The last metal fragments and labels must be removed in subsequent processing phases, particularly in the pre-spinning operations. Fibres recycled in this manner can also be utilised in open-end yarn spinning. Because the fibres produced by this technique are a little shorter, more virgin fibres must be added to obtain the same yarn quality as post-consumer jeans legs alone. In addition to this industrial process, several Internet sources explain the transformation of jeans materials into handbags, toys, carpets, quilts and patchwork. You may find a large variety of these products, as well as instructions on how to build them, at. Although this type of recycling might be thought of as high-end recycling that adds value, the economic worth of such activities is very minimal, just a small fraction of all wasted jeans can be reused in this way.

3 Conclusion

Securing our planet and sustainable fashion has become a vogue word today. As we already know the giant denim industry produces boundless quantities of second hand trash, or unsalable stock that results in wide environment pollution in itself. Today, the sustainable development of products is of main focus. Sustainability cannot be accomplished quickly. It wants cautious analysis and understanding. It should also be applied at each and every stage of the supply chain. The requirement for sustainable production needs to be communicated to manufacturers: likewise, sustainable use and disposal ought to be communicated to the buyer. Recycling, upcycling, use of substances that doesn't hurt the atmosphere, and water and energy savings are the key parts of sustainable denim production. Some leading makers and brands of jeans have already begun analysis into and implementation of sustainable production. Sustainable practices ought to be grownup alongside the fibre, spun into the yarn, woven alongside fabric and imparted whilst colouring and finishing. Above all, client awareness relating to the sustainable use and disposal of denim is very important.

References

1. Periyasamy AP, Duraisamy G (2018) Carbon footprint on denim manufacturing. In: Handbook of Ecomaterials. Springer International Publishing, Cham, pp 1–18. https://doi.org/10.1007/978-3-319-48281-1_112-1
2. Levi Strauss & CO (2013) The life cycle of a jean
3. http://levistrauss.com/wpcontent/uploads/2015/03/Full-LCA-Results-Deck-FINAL.pdf
4. Khalil E (2015) Sustainable and ecological finishing technology for denim jeans, department of textile engineering. AASCIT Commun 2(5) (World University of Bangladesh, Dhaka, Bangladesh)
5. Shalini N (2008) Fabric and garment finishing: basic washes in denim fabric, Fibre2fashion. https://www.fibre2fashion.com/industry-article/3735/fabric-and-garment-finishing-basic-washes-in-denim-fabric
6. Alam MR, Islam T, Rahman M, Antor MAAN, Rahman R, Tamanna TA (2020) Sustainable denim fabric washing with post-used rubber shoe sole: an eco-friendly alternative of Pumice stone. Indian J Sci Technol 13(48):4723–4731. doi:https://doi.org/10.17485/IJST/v13i48.1974
7. Hoque MS, Rashid MA, Chowdhury S, Chakraborty A, Haque AN (2018) Alternative washing of cotton denim fabrics by natural agents. Am J Environ Prot 7(6):79–83. https://doi.org/10.11648/j.ajep.20180706.12
8. Csanak E., (2014) Sustainable concepts and eco-friendly technologies in the Denim industry. In: International conference on design and light industry technologies. Óbuda University, Hungary
9. Radhakrishnan S (2016) Denim recycling, textile and clothing sustainability. In: Muthu SS (ed) Textile science and clothing technology series, pp 79–125. https://doi.org/10.1007/978-981-10-2146-6_3
10. Periyasamy AP, Wiener J, Militky J (2017) Life-cycle assessment of denim from Sustainability in Denim. In: The textile institute book series, pp 83–110. https://doi.org/10.1016/B978-0-08-102043-2.00004-6
10. Pal H, Chatterjee KN, Sharma D (2017) Water footprint of denim industry. In: The textile institute book series, pp 111–123. https://doi.org/10.1016/b978-0-08-102043-2.00005-8
12. Kan CW (2015) Washing techniques for denim jeans. In: Paul R (ed) Denim manufacture, finishing and applications. Woodhead publishing series in textile, pp 313–353. https://doi.org/10.1016/B978-0-85709-843-6.00016-0
13. Specter F (2019) Why you should never wash your jeans, yahoo lifestyle. https://au.lifestyle.yahoo.com/never-wash-jeans-000344119.html. Accessed 02 09 2021

Implications of Sustainability on Textile Fibres and Wet Processing, Barriers in Implementation

Srivani Thadepalli and Shreyasi Roy

Abstract Fashion industry is heavily dependent on textiles. Without textiles, there would be no fashion apparel or home furnishings. With the world and humanity now concerned with the dire problem of waste disposal caused by fast fashion lifestyle of the generation Z, the first steps towards a viable solution have to come from the raw materials used and process of production of these fashion items—be it apparels, accessories, home décor or packaging material. The textile processing sector contributes to a huge generation of waste and pollution to the environment. To reduce the consumption of water and chemicals several methods of finishing such as thermal dry, foam and spray techniques are developed. The objective of this chapter is to investigate the eco-friendly options available in terms of machines, technologies, chemicals and processes to make the textile sector environmentally sustainable. The areas mainly reviewed under this study include sustainability aspects involved in dyeing and printing using natural dyes and need of research and development in sustainable ways of using synthetic dyes and pigments. The study also delves into the various technological innovations for sustainable processing and zero water finishing including Ozone treatment, Plasma, Micro encapsulation, Nano and Laser Technology, etc. The chapter emphasises the reasons which inhibit Indian textile industries from embracing sustainable processing and suggestions to make sustainability mainstream in Indian textile industries. Sustainability and fashion are both the driving forces for these revolutionary innovations to be implemented, which contribute to greener textile factories and goods that are more environmentally friendly.

Keywords Sustainability · Eco-friendly processing · Textile finishing · Sustainable fibres · Technological Innovations

S. Thadepalli (✉)
Associate Professor, DFT, National Institute of Fashion Technology, Hyderabad, India
e-mail: thadepalli.srivani@nift.ac.in

S. Roy
Student, FMS, National Institute of Fashion Technology, Hyderabad, India
e-mail: shreyasi.roy@nift.ac.in

© The Author(s), under exclusive license to Springer Nature Singapore Pte Ltd. 2022
S. S. Muthu (ed.), *Sustainable Approaches in Textiles and Fashion*,
Sustainable Textiles: Production, Processing, Manufacturing & Chemistry,
https://doi.org/10.1007/978-981-19-0538-4_6

1 Introduction

The sustainable development is defined as the 'development that meets the needs of the present without compromising the ability of future generations to meet their own needs' by the UN World Commission [1]. The current prevailing definition refers to sustainability as a 'dynamic equilibrium in the process of interaction between a population and the carrying capacity of its environment such that the population develops to express its full potential without producing irreversible, adverse effects on the carrying capacity of the environment upon which it depends' [2].

Referring to textiles, 'Sustainability' can be expressed as using environment friendly and refined methods while producing fabrics, which mean establishment of thoughtful practices that, would conserve energy, natural resources and reduce negative social, economic and environmental impacts. A 'sustainable fabric' is said to be produced in ways which minimise effects on environment through choosing of practices such as recycling of water, recycling and recovery of raw materials and heat from steam generation plants and wastewater, decreasing the use of dyes, chemicals, and replacement of harmful chemicals with safe and non-hazardous substances.

Sustainability begins with the very conception of the product, which considers the aspects like choice of fibre used as raw material, type of yarn, fabric, dyes used, methods of fabric finishing and garment processing. In any industry, manufacturers would enjoy benefits if sustainable practices are employed such as enriched brand reputation and customer satisfaction among the garment producers, retailers, and consumers, who are concerned with safe environment. Reduction in costs through lessened use of dyes, chemicals, energy and declined wastewater treatment costs upholds business ethics and practices.

A single textile unit can use up to 600 gallons of water and more than 200 chemicals to make finished fabrics from raw materials [3]. The production of fabrics also results in solid wastes like sludge, scrap material, oily cloths, etc. According to a report by World Bank, about 17–20% of all contaminants in water are by-products of textile dyeing and finishing processes [4]. Not only production of textiles cause pollution, but these products keep releasing heavy metals, toxic effluents after it is discarded and ends in the landfills and waterbodies. Sustainability and sustainable approaches are necessities of the fashion industry at present.

With more than 400 billion square metres of annual production of textiles [5], this leads to 60 billion square metres of textile wastage in the cutting rooms. As a leading producer of textiles in the world, India as a producer must be conscious of the toll of production on environment and the people associated with the productions. While the western consumers have become conscious of over consumption and sustainability, Indian producers, not so much. The textile industry in India consumes resources rapidly and releases more waste than what the environment can degrade. Statistics of the fashion industry shows that

- Fashion and textile industries are the largest polluters of the planet next to oil [6]
- Annually, 10% of carbon emissions are released by fashion and textile industry [7]

- Fashion industry causes 20% of industrial water pollution worldwide [8]
- Only 8% is reused, 10% is recycled, 25% is incinerated, 57% ends up in landfill [9].

The green economics (managing the relationships between people and environment that are interconnected), green marketing (referring to the procedure of developing products and advertising based on their observed or real environmental sustainability) and impact investing (Creation of social or environmental benefits in addition to economic gains) have become the points of focus for companies today satisfying the Triple Bottom Line approach [10].

For a good reason, sustainability is a burning topic today amid producers and consumers alike. The environmental and social impacts of fashion industry from emission of greenhouse gases, loss of biodiversity and existence of micro-plastics to the exploitation of garment workers and systematic racial injustices can no longer be put out of sight. Sustainability became a buzzword across the industry, however mere talking is not enough. In the name of green, many natural dye manufacturers, practitioners and natural fibre suppliers have created enormous surge for the sustainable products, increased consumer demand through various platforms of advertising and promotion. Many NGOs and entrepreneurs are quite unreasonable in pricing being business minded. Much awareness is required about the extent of sustainability of natural materials with relevance to the implications of their availability for the population. The chapter calls for discussion on various practical aspects involved and further barriers to implementation of sustainability in Indian textile industry. A special focus was given to the aspects of sustainability of natural dyes and organic cotton in today's context.

This chapter provides a detailed note on implications of sustainability involved in textile chain right from selection of natural fibres and use of natural dyes to various aspects of sustainability involved in wet processing. It conveys information on various safe and eco-friendly options available in terms of machines, technologies, chemicals and processes in wet processing sector. The end of the chapter emphasises the reasons which inhibit Indian textile industries from embracing sustainable processing and suggestions to make sustainability mainstream in Indian textile industries.

2 Approach to Selection of Sustainable Fibres

The world of textiles is based on the availability of raw material—fibres, which connect us to many issues of water scarcity, climate change and waste generation. The world's food, fuel and fresh water supplies are coming under increasing pressure due to the rising temperatures, population growth and increasing levels of per capita consumption. Textile has always formed a substantial part of sustainability and innovation.

The natural plant and animal fibres-cotton, silk and wool, which were once considered as solutions for sustainability based on their biodegradability and renewability, are no longer able to serve the purpose due to raw material crisis and also a diversity of ethical concerns and animal welfare existing. Civilisation and the social status of the wearer, the culture, the fashion trends have contributed towards innovative changes in the clothing style, and in turn in textiles manufacture.

As cotton crop takes a gigantic load of herbicides and pesticides at the cost of human health risks, it is not the answer to fulfil the increasing demand of clothing in the world. Organic cotton, Hemp, Ramie, Linen and organic Bamboo that is mechanically processed are inclusive options of today's textile and garment industries. Innovations in textiles have brought alternative plant-based regenerated fibres of first, second and third generation (Viscose bamboo, Modal & Lyocell) into the spotlight and as a replacement to their natural synthetic counterparts. Ecovero by Lenzing™ is a new type of Viscose produced with very low emissions. The 'zero-waste' utilisation of the wood-based cellulosic fibres, the fibres of Okra plant and Agave Americana are produced utilising the agricultural waste/biomass would enable a viable and sustainable industry.

The recycled wool, the vegan wool developed from the enzymes of mushrooms, Ecolife responsible Leather and Down, Peace/ Ahimsa silk derived from hatched cocoons are considered as sustainable animal fibre options. Pinatex, Apple skin/Pallemela and Orange the by-products of fruit industry and Cork fabric out of tree bark and Green Cactus leaves have crossed the verge of sustainable fashion world as 'cruelty free Eco-Vegan Leather'.

Increasing use of synthetic textile fibres makes a substantial impact on the environment, due to persistent micro-plastics. Recycling textiles/synthetic fibres is a sustainable and environment friendly practice. Synthetic waste from ocean plastic, abandoned fishing nets, waste fabric, etc. are recycled using closed-loop system to produce 'Econyl', which is a recycled nylon.

We are amidst an ecological and environmental crisis and in need for the creation of awareness towards sustainable approaches towards selection of fibres as well as processing methods further is a matter of urgency.

2.1 Organic Cotton—Is It a Sustainable Option?

Cotton that is grown without undue use of pesticide, insecticide or chemicals and fertilisers is known as Organic Cotton [11]. The methods of organic cotton farming ensure soil health, fertility, reduce chemical effluents in the groundwater and do not use genetically modified seeds for plantation and production. Benefits of organic cotton include

- Organic cotton farming produces less environmental footprint than conventional cotton farming. It uses 88% less water, 62% less energy.
- The rearing and production use an amalgamation of scientific methods and traditional farming. It also keeps the workers and farmers safe since there are no chemicals and pesticides in use.

- Organic cotton products also help sustain the livelihoods. When a consumer purchases organic cotton products, they are investing in the environment, the farmers, themselves and the future [12].

While organic cotton is a sustainable option, there are a few drawbacks that should be noted. Unsustainable farming practises may eclipse the potential benefits of organic cotton [13]. Large corporations entering the field of cotton cultivation have increased such practises. The drawbacks of organic cotton are

- **Lower yield**: Since organic cotton does not use any chemical fertilisers or pesticides, it has a lower yield. More land is required for cultivation of organic cotton to match the equivalent of the conventional yield.
- **Water consumption**: On an average, organic cotton has lower consumption of water per tree. However, to match the yield of conventional cotton, more land has to be used for cultivation, hence, more water is used for farming. Thus, water consumption is a complicated metric in case of organic cotton.
- **Ethical**: Despite the benefits of organic cotton, consumers must be aware of the source of the same before purchasing. Sustainable option does not always equate to ethical. Since large corporations are in the market as producers of organic cotton, consumers must be aware and educate themselves on identifying the original organic cotton and not be fooled by Green Washing [14].

Hence, in today's market there is absolute need for sustainable alternatives in the market for organic cotton like Hemp as it is high yielding, requiring much less water than cotton and Tencel, a regenerated fibre produced in closed loop. However, they are not currently produced in the mass scale that is needed to completely replace the use of cotton.

3 Sustainability in Textile Wet Processing

Textile wet processing, which includes preparatory, colouration and finishing processes, is an important part of production and manufacturing. Wet processing is done as a means of value addition to textiles as it enhances the aesthetics, comfort and tactile properties. However, an enormous amount of water is used in wet processing operations that get contaminated with combination of chemicals discharged as effluent, which further makes it difficult to treat or biodegrade. Textile industry is thus water, chemical and energy intensive industry, which puts a lot of strain on global resources [15].

3.1 Introduction to Sustainable Practices in the Colouring Industry

Dyeing and Printing are the major colouring options for textiles, which utilises synthetic dyes and pigments. Many chemical-based ingredients used in the application processes of synthetic dyes are found to cause major health hazards apart from resulting large-scale pollution in the waterbodies.

Natural dyes extracted from organic sources like plants, animals, minerals and other natural sources are easy to extract, cause less pollution, non-hazardous to health, negligible impact of effluent released, renewable and also known to offer protection against UV rays [16].

Despite these benefits, natural dyes are not widely used referring to their requirement of mordents, moderate colourfastness, selective suitability on fibres, difficulty in standardisation and reproduction of shades besides being expensive and time consuming. In India, Alps Industries in Ghaziabad initially has done extensive research on production of natural dyes in industrial scale, later, Ama Herbals of Lucknow and Bio Dye of Goa also joined the efforts to produce concentrated dye powders of natural dyes in pilot scale to test the possibility of mass scale production for industrial applications.

3.2 Using Natural Dyes—Is It a Sustainable Option?

Though Natural dyes are a best option for fabrics, currently only one percent of textiles are dyed using natural dye sources. Indian dyestuff industry currently produces about 300 thousand metric tonnes of dye annually [17]. However, the statistics show the sheer volume of synthetic dyes and pigments required to feed the current fashion demands causes a major environmental degradation. Usage of natural dyes is mainly limited to cottage industry and some handloom sectors. However, natural dyes are not a mainstream option because of sustainability issues.

Most of the natural dyes have very less dye content in the raw source. To meet the huge requirement, the agricultural land used to cultivate food grains to feed the huge population of the country may have to be compromised. Plant-based dye sources therefore do not have much spare land to be cultivated on. Dye stuff from animal sources requires infrastructure and initial setups to have a mass population of the animals before dye can be extracted from them. The investment is costly and the current farmers are not economically stable. Thus, they are not much interested to produce alternate dye sources where they can produce cash crops and genetically modified food grains with higher yield. Additionally, natural sources require time to mature before dyes can be extracted, e.g. trees need to grow from saplings before they can be processed.

This also brings up the issue of indiscriminate deforestation. To meet the demands of the market, the existing plant and animal life maybe indiscriminately decimated to produce dyes. This way, sustainability cannot be achieved as the demands of the current generation are met compromising the requirements of future generations. In this respect, natural dyes are far behind as the industry does do not have the raw material source, capacity or capability to meet the current demands of the market.

Though natural dyes and natural pigments are definitely the only environment-friendly option for textile dyeing and printing, for the reasons mentioned above, more research to come up for the use of synthetics dyes and pigments in a sustainable way, which is the only long-term feasible solution besides research and development on industrial production and application of natural dyes.

3.3 Growth of Indian Dyestuff Industry

Indian Dyestuff industry currently produces 300 thousand metric tonnes of dye annually [18]. The Indian dyestuff industry is one of the principal chemical industries in India, the second-highest export segment in chemical industry and also an important economic sector for progress of our country (Fig. 1). Maharashtra and Gujarat represent 90% of dyestuff production in India. Almost 80% of the total dyestuff produced is consumed by the textile sector annually based on high demand for polyester and cotton, globally.

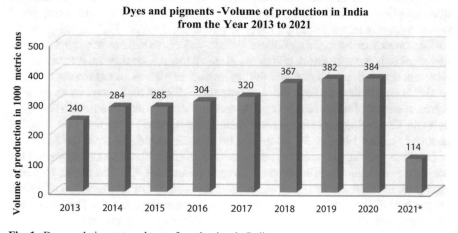

Fig. 1 Dyes and pigments-volume of production in India

3.4   Implications of Sustainability on Dyeing and Printing Industry

Before colouration, the cotton fabric undergoes various preparation processes such as de-sizing, scouring, bleaching and mercerization which make the fabric absorbent and ready for dyeing, printing and finishing. Various chemicals including strong acids and alkalis are used in preparation, which also consume abundant water. Washing off of unfixed chemicals and other auxiliaries into effluent poses serious environmental concerns. Hence, identifying alternative eco-friendly processes for textile wet processing is of vital interest [19].

Bio Processing using Enzymes is the new norm followed by industries. De-sizing and bio-scouring of cotton and micropoly fabrics using thermostable xylanase from *Bacillus pumilus* indicated good de-sizing efficiency, lowered wetting time of fabrics, higher whiteness values [20]. Cost reduction through better utilisation of spent water, chemicals and energy in the process and subsequent decrease in the resultant wastewater can be achieved by optimisation of scouring parameters [21]. The leftover water from the rinsing process of H_2O_2/peroxide -bleached fabrics can be reused for dyeing by treating with immobilised enzymes. The new one bath process of de-sizing (*amylo-glucosidase*), bleaching (*glucose oxidase*), decomposition of hydrogen peroxide (*catalase*) and dyeing (reactive dyes) of cotton fabrics using enzymes is found to produce comparable results. Compared to the conventional de-sizing, scouring, bleaching and dyeing sequence, the one bath process is a sustainable dyeing option as it reduces the demand for auxiliaries, coupled with energy and water savings there by achieving lower environmental impact [22]. In reducing the surface tension of mercerizing liquors, ordinary alcohols such as isopropyl alcohol, methanol or butanol can work efficiently at much lower cost; they are renewable and also vaporise during yarn or fabric drying having no environmental impact.

Salt is used to develop negative charge on the surface of the cotton in conventional cotton dyeing. **ColorZen technology** focuses on creating a natural attraction between dye and fibre using a quaternary ammonium compound to permanently attach a positively charged amino site on the cellulose molecule. Right from the field, raw cotton fibres get treated after the seeds are removed and later spun into yarn. This pre-treatment enhances exhaustion of the dyes faster, uses 90% less water, 75% less energy and 90% fewer auxiliary chemicals. It also cuts out consumption of dye to almost half of those processes that necessitate salts in the dye bath. Huntsman had developed a new line of dyes for cotton called **Avitera**, which has bonds to get fixed to the fibre more readily. Three reactive groups are attached to the colour-providing molecule or dye formula's chromophore, compared to one or two reactive groups common for cotton dyes. Avitera dyes require 1/4–1/3rd less water and 1/3rd less energy and the total dyeing process takes about four hours in place of seven hours required by conventional dyes [23]. **Pre-treatment of cotton fabrics with cationic starch** before dyeing with reactive dyes by a continuous dyeing method has been found to give level dyeing without the presence of salt. Marked improvements were observed even in the dye fixation compared to untreated cotton fabrics showing good fastness to washing and rubbing [24]. In dyeing cotton with reactive dyes, use of **bio-salt tri-sodium citrate** instead of sodium chloride has shown higher dye uptake values and lower TDS content in spent liquor [25]. **Organic salts** such as magnesium acetate, tetra sodium edate and sodium salts of poly-carboxylic acids besides being **biodegradable** can minimise salt and water consumption when used as fixation and exhaustion agents [19].

Though indigo dye was originally extracted from the indigo plant, most of today's indigo pigment is chemically synthesised. Being insoluble in water, Indigo dyes must undergo a process called chemical reduction which uses sodium hydrosulphite, a salt that is corrosive and ultimately end up in the waterways. According to German chemical supplier DyStar, 70,000 metric tons of indigo are used for denim products every year that is quite alarming. DyStar came up with a Cadira Denim System that replaced hydrosulphite with Sera Con C-RDA an **organic reducing agent**, and combined with 40% solution of DyStar Indigo Vat to create the 'cleanest indigo on the market', what it commercially called. The Korea Advanced Institute of Science and Technology has invented a process of producing indigo dye from bacteria. The scientists metabolically engineered Corynebacterium glutamicum, which produces an indigoidine, a blue dye that is natural and more sustainable than conventional indigo. White cotton fabrics dyed using indigoidine are tested for further confirmation that it has the potential to replace indigo [26]. Advance Denim introduced its newest dyeing innovation, BioBlue Indigo (**eco-friendly replacement for sodium hydrosulphite**) for reducing indigo in a sustainable manner resulting in lower COD and BOD levels. The revolutionary Bigbox dyeing method saves up to 98% of water necessary to dye indigo yet creating same deep and consistent shades. By implementation of **RO recycling systems**, denim industries can reduce water consumption by 58% and will be able to recycle 100% of the waste water used in the denim finishing processes [27]. The sludge extract of wild berry, *Bokbunja* can be a safe alternative to sodium dithionate. *Bokbunja* mainly used for producing traditional wine contains large amounts of anthocyans, sugars and phenolic components. The water extract of the sludge at its 3% concentration is efficient in reducing indigo [28]. Monosaccharides such as glucose, fructose and galactose and reducing disaccharides, viz., lactose and maltose along with sodium hydroxide can be used as **green reducing agents** for preparation of stock vat to reduce consumption of $Na_2S_2O_4$ [29].

Pre-reduced liquid colourants developed by dye suppliers provide remarkable environmental advantages. DyStar reported that the production process of natural Indigo dyestuff in total inclusive of land requirements, harvesting, dye extraction, and residual biomass is less environmentally favourable than **optimised synthesis of indigo** [30]. Thiourea dioxide is an alternative to sodium hydrosulphite, which is more environmentally friendly because of its lower toxicity and higher stability [31]. A same level of reduction as sodium hydrosulphite was found by using Molasses. [32]. A recent innovation of using **hydrogen gas for reduction** of sulphur dyes not only eliminates the load of reducing agent but also allows shipping of pre-reduced dye at concentrations as high as 40%. The higher reduction potential can result in superior wash fastness [33].

Through innovations in process design, Indigo Blue of North Augusta has pioneered in sulphur dyeing using energy-efficient methods that allow dyeing with sulphur black dye at a temperature as low as 30 °C. KG Denim, Coimbatore, India has become a leader in sustainability with regard to denim fabric and is appealing to offer the most sustainable denim in the world as 94% of the process water that leaves dyeing gets recovered and returns to manufacturing. 75% of the electrical energy used is produced by wind farms, and the consumption of energy has been reduced by the use of energy-efficient motors, improved internal lighting and through implementation of an energy monitoring system. A system of pre humidification of fibres was found to intensely reduce water consumption. Cotton waste amounting to nearly 10% of the raw cotton arriving factory is recycled by the company for use as fuel for boilers, replacing oil and coal that are non-renewable [34]. To reduce water usage and increase energy efficiency, modern wash boxes with counterflow technology, which allows the movement of yarn and wash liquor in opposite directions [35]. A conventional method that is prevalent for recovery of indigo is to precipitate indigo dye by adding aluminium sulphate to the wash liquor and then dissolving the precipitate to release the dye. Another approach to recovery of indigo is employing palygorskite clay, which absorbs the indigo followed by conversion of the recovery by-products into Maya Blue, an organic–inorganic hybrid pigment having suitable applications in the painting and coating industry [36].

Dye houses can adopt the use of Engineered microbes that produce stable colours. First, the textile fibre is embedded with microbes added to the dye solution and then the microbes are allowed to grow in fabric with the supply of nutrients given. At an identified stage of growth, heat is applied to break the membranes of organisms resulting in the colour to get chemically attached to the fibres with the support of metal ions and salts present in the cytoplasm of microbes. This process is considered very efficient as it requires only a single final rinse guaranteeing 90% savings of water, and 20% of savings on energy over standard processes. In the process of exploring eco-friendly alternatives to artificial dyes, the Dutch fashion project Living Colour has developed a method to dye textiles using bacteria without using any hazardous chemicals [37]. Using dye suspension instead of dye solution to dye, the fabric reduces the effluent. It can be easily filtered from water, thus reducing the environmental impact.

Decreased use of chemical auxiliaries: In addition to indigo, sodium hydroxide and sodium hydrosulphite, majority of operations in denim manufacture uses wetting agents, sequestering agents, dispersants, softeners, fixatives, oxidisers, acids, sulphur dyes and sulphur dye reducers. These chemical auxiliaries can contribute to pollution as they are generally from non-renewable sources. Many a time, they are used in unnecessary quantities with minimal advantage but add to the dyeing costs. For reducing the toxicity of effluent and for curtailing consumption of water, chemicals and energy other advanced and innovative approaches like low-liquor continuous processing, microwave-assisted processing, irradiation technology, use of Nano and biomaterials for value addition, foam finishing and digital printing have been developed [38]. The use of sustainable natural mordents, thickening agents and use of natural agents like madder and jiggery for reducing indigo in printing are gaining prominence [39].

Digital Printing—The major sustainable option at present: Digital printing technologies offer a way forward towards clean, efficient, profitable manufacturing, being more sustainable by nature than traditional techniques. Year 2018 witnessed a worldwide water savings above 40 billion litres by digital printing. The usage of water in conventional rotary screen printing is 50–60 L/m, whereas digital printing usually consumes lesser quantities to the extent of only 10 L/m of cloth. Digital printing uses only 10% of the total volume of water used by screen printing as it requires no water for fixation. The discharge of noxious effluent and the consumption of water can be eliminated by the use of digital printing with pigments having low volumes of liquid dispersions. It offers a positive environmental impact as there is no need of washing during post-processing since heat fixation alone can achieve required colour fastness as against lengthy processes of steam fixation and washing off procedures [40]. The new generation of digital printing presses offers the benefit of printing even on ready-made garments. The latest development exhibited in the Promotional Products Tradeshow of Industrial textiles is issuing of an A2 fire protection certificate for both the material and ink used in high-quality digital prints on glass fibre, which offers fire protection [41]. The process of digital printing on polyester can be done by a simple and fast two step sublimation process with negligible usage of water. Hence, the technology is more suitable for the fast-fashion retailers such as H&M and Zara which prefer fast delivery schedules rather than the traditional models in which the delivery schedules run up to several months [42].

Sensient released printing inks and processes that are certified by OEKO-TEX. The main aim is to achieve the highest standards of quality of printing and performance besides saving total consumption of water, energy and reduction in maintenance of waste. Roland DG developed **environmentally friendly inks**, developed software for maximum efficiency of inks to ensure the consumption of inks only to the necessary level. It also shifted from using rigid plastic to **flexible foil cartridges for ink.** Kornit digital offered an eco-friendly 100% waterless printing system using **biodegradable inks** which require no pre-treatment, steaming or washing. Epson developed a new sublimation printer for printing with fluorescent inks/ Neon colours for sports application which are wash fast and breathable.

4 Textile Finishing—Processes and Implications

Finishing is what makes the fabrics suitable for intended use. Finishing plays a major role in textile industry as they leave a profound effect on fabrics that are treated using mechanical or chemical processes. They have commercial importance as they can increase the acceptability of the final product through its aesthetic or functional performance. It includes all processes except dyeing and printing. The objective of various finishing processes is to make woven and knitted fabrics saleable and suitable for various end uses [43]. Unlike mechanical finishes, chemical finishes are extremely invasive of various chemicals. The effluents released in the environment are extremely harmful. Wrinkle-free finishes require formaldehyde, a well-known carcinogenic. The softening finishes applied to casual wear yoga pants contain siloxanes, synthetic compounds which build up in the environment for being non-biodegradable. All these chemicals do affect the workers causing various skin diseases, lung diseases and more. They also affect the consumer, resulting in health hazards like dermatitis [44].

Environmental considerations and sustainable methods have inspired the development of eco-friendly, low wastage finishing processes of textiles. Great steps are being taken and innovations are being made to reduce the carbon footprint of the textile industry and to replace the harmful conventional methods of finishing of textiles. The most prominent and emerging finishing method is the zero water finishing.

4.1 Zero Water Finishes on Textiles

4.1.1 Ozone Treatment as a Sustainable Option

Ozone, made up of 3 oxygen atoms, is an excellent oxidising agent, which can be applied both as a gaseous or aqueous agent [45]. Ozonisation is a 'green' process since it does not require steam or water. It highly reduces time, costs, resources, water consumption and also reduces effluents as Ozone does not produce any waste [46]. Ozonisation is also useful as a pre-treatment of raw cotton; the high oxidative properties of ozone can successfully bleach the cotton fibres during processing.

Ozonisation in Denim industry: Ozone treatment is an alternative bleaching method in denim processing, which is a great advancement in the garment sector as it plays a vital role in reducing environmental impact. Traditional denim wet processes include several toxic and polluting chemicals which are released in the environment as wastage. In the process of ozonisation, O-oxygen is converted to ozone in the machine, dampened denim products are exposed to Ozone and then it is reconverted to ordinary oxygen before being released. While chemical bleaching and stone washing requires 6–7 washes, ozonisation requires only 2–3 washes to achieve the same results.

Studies on effect of Ozone: Different ozone applications to natural dyed cotton fabrics produced different shades and effects in the dyed fabric [47]. Ozonised samples had high colour fastness and rubbing fastness except light fastness. Discharge printing can be done using ozone gas [48]. Ozone also increases hydrophilic properties of synthetic fibres.

Despite its usefulness, ozone has a few drawbacks like yellowing of the fabric after ozone treatment, loss of tensile strength in fabrics if the right process is not followed, corrosion and damage of metal and plastic parts of machine due to high oxidative property of ozone. Onsite production of ozone is necessary as it gets decomposed when stored, which entails high capital investment for initial installation and subsequent maintenance.

4.1.2 Plasma Treatments

Plasma treatment, being a dry finishing process, requires only gaseous interaction with textile materials. Thus, it removes the requirement of water for processing which results in no effluents. Plasma treatment uses non-toxic gases for treatment like oxygen, nitrogen or inert gases like argon [49]. These gases are completely eco-friendly and do not generate any pollutant during processing. Plasma treatment is extremely versatile, carried out at room temperatures. It removes any need of heat sources, washing, drying steps thus reducing time and resource requirement.

Plasma treatments are used for surface modifications and bulk property enhancement of textiles and preparatory processes of both natural fibres (cotton and wool) and man-made textiles (rayon and polyester). Studies show that plasma treatment improves dye ability of polymers and colour fastness of textiles. Plasma can be applied onto the textile surface via spray method also, in which the textile surface is coated with tiny droplets of liquid plasma. These droplets are flattened due to heat transfer to the textile surface enabling the droplets to be cooled off, flattened, shrunk and solidified on the textile surface as a continuous film. It is observed that the specific surface area of cotton increases after oxygen plasma treatment. Plasma treatment on wool achieves shrink-resistance [50] and also reduces felting [51].

Oxygen plasma is used to improve the softness of cotton and other cellulose-based fibres. Choloromethyl dimethylsilane plasma finishing is applied to rayon textiles to produce anti-static surface. Cotton and polyethylene textiles can attain hydrophobic property when finished with Siloxan and perflurocarbon plasma treatment. Argon plasma treatment improved dye ability of polyamide textiles. Plasma treatment setup can be made vertically as well, which optimises floor space. Though the initial set-up may be cost intensive, it is very efficient in the long run with increased productivity, decreased time and elimination of chemical and water resources. Plasma treatment is definitely a sustainable finishing method.

4.1.3 Super Critical Carbon Di-Oxide

Netherlands-based DyeCOO Textile Systems have developed an innovative dyeing technology which uses super critical carbon dioxide for waterless dyeing method widely known for textile processes like de-sizing, scouring and application of different finishes. The supercritical carbon has high solvent power and easily dissolves dye, resulting in higher dye penetration in the fabrics. The unreacted dye remains as the dye powder which can be easily extracted and reused. This is a cost effective and time-saving method as no further fabric treatment is required [52]. Carbon dioxide the raw material used is non-toxic, non-flammable, inert, inexpensive and easily available. The super critical conditions are easily achieved; by simply changing the pressure, the density of the solvent can be managed [53]. Compared to conventional methods, super critical carbon treatment requires only a fraction of energy and time. It is fast, efficient, highly productive process with no resultant waste and effluent in the processing to worry about.

4.1.4 Laser Technique

The increasing concerns of pollution and environment protection established the need for environmentally friendly methods for surface treatments. Laser treatment enables precise surface modification in a short time based on physical principles. It offers environmentally clean, easy to apply/control with zero consumption of water or chemicals proving several advantages over conventional methods of chemical finishing [54]. Though lasers are known for their cutting capabilities, they can also achieve a wide range of finishing effects which would otherwise need multiple chemical and mechanical finishing processes, using different machines to produce making laser technique eco-friendly and versatile. A gaseous mixture of carbon dioxide, helium, hydrogen and nitrogen is used for laser treatment. Electrical current when passed through this gaseous mixture in a mirrored tube, the gases produce thermal energy which is reflected by mirrors and intensified to produce a single beam. The laser beam is then directed on the fabric surface using highly mobile motorised mirror system [55]. The properties of fabric hand, namely, smoothness, softness, stiffness, wrinkle recovery and drape-ability could be changed successfully through laser treatment [56].

4.2 Finishes with Reduced Water Consumption

4.2.1 Nanotechnology

Application of Nanoparticles on textiles to produce a variety of functional finishes on textiles is becoming very popular. Nanotech finishes can be applied on natural and synthetic fibres to create easy care fabrics like stain resistant, wrinkle resistant and hydrophobic properties. Nanoparticles of antimony pentoxide produce flame retardant finishes, Montmorillonite produces UV resistant finish and Silver protects fabrics from dust and stain, thus providing stain resistant finish for suits [57].

Application of nanomaterials: Nanoparticles can be applied on the textile surface via foam (savings on water), spray coating, vapour deposition, electrospinning (nanoparticle-induced solvent is passed through an electric field where the solvent evaporates and the nanopolymer forms elongated fibres which weave together in a mesh form) [58]. Nanotechnology enhanced textiles give us various functional performance like dust resistance, wrinkle resistance, water resistance (lotus effect) self-cleaning fabrics and UV resistant. Nano-applications are mostly used in technical textiles like wound dressing.

4.2.2 Microencapsulation

Microencapsulation is the process of coating a droplet of chemical to form a capsule. Microcapsules get fixed in the fibres when coated to the fabric surface, which provide various finishes in textiles [59] like phase changing thermos-regulatory microcapsules that change phase from solid to liquid depending on the weather and temperatures, capsules of fragrant components like lavender and other aromatic essences for odour control and antimicrobial micro capsules that prevent bacterial infection, used in medical textiles.

While nanotechnology in textile finishing is a mostly efficient sustainable method for using less water based on its suspension mechanism, however it is not completely eco-friendly as assumed. Nanomaterial and microcapsules, which coat the textile surface for the desired functional finishing, start to leave fabric after certain number of washes. The discharge of nanomaterials in the environment, due to usage and abrasion while washing, may appear negligent, however, over time; the build-up in the ecosystem is huge. The bonding between the textile and the nanomaterials deteriorates throughout the lifecycle of the product, from the inception to disposal [60]. Despite the nanomaterials and microcapsules are photocatalytic, they become highly reactive to other materials while being discharged during the lifecycle of the product.

5 Technological Innovations

As Industry 4.0 has made its advent into the current civilization, smart tech powered by artificial intelligence is making its way in all the industrial sectors. Textile industry is also not far behind. Keeping with the times, here are a few companies who developed state-of-the-art technological innovations for textile finishing and processing [61]. Modern technological advancements in textile processing have increased efficiency and reduced pollution. These tech advancements hugely reduce resource consumption and waste generation (Table 1).

6 Reasons Which Inhibit Indian Textile Industries from Embracing Sustainable Processing

Sustainability and sustainable business practices have driven the market and firms towards the triple bottom line framework, i.e. good for the people, business and also for the environment. Indian textile industry is mainly comprised non-integrated enterprises such as small-scale spinning, weaving and finishing. The distinguishing structure of Indian textile industry is the result of the prevalent government policies which encouraged labour-intensive small-scale enterprises and discriminated against large-scale firms. In contrast to other global textile producers, India's small-scale textile enterprises mostly use outdated technologies. Indian textile industries are classified into several categories as given [62]:

Table 1 State-of-the-art technological innovations for sustainability in processing sector

Innovation	Developed by	Principle of functioning
TexCoat 4G	Baldwin Technology & Co, St. Louis	This non-contact spray applicator with efficient nozzles can ensure uniform application of finishes either on one or both sides of the textile in the same time. Smart technology backed by artificial intelligence switch the overlapping nozzles on or off as required to avoid overlapping of spray application. Completely enclosed unit, the spray applicator has zero wastage of finishing chemical
Tempacta	Benninger AG, Switzerland	A washing steamer unit developed for low-tension washing process for knitwear, used for diffusion washing. The unit continuously measures online the amount of contamination in the water and exactly regulates freshwater requirement to counter it, guarantees the lowest utilisation of water and energy consumption, besides ensuring reproducibility of washing results
MonforClean module	A. Monforts Textilmaschinen, Germany	An exhaust air system incorporated in a tenter frame. The heat released as a waste is used in preheating the drying air to reduce the requirement of conventional heat sources
Knit Merc	Goller TextilemaschinenGermany	This machine is designed for dry-on-wet mercerization of cotton and other cellulosic knitted fabrics. Knit Merc uses the lowest possible tension with lesser than 3% variation in dimensional stability. The machine can accommodate 8.4 m of fabric in the compartment as well as 4 m in the first chain section. The production speed of Knit Merc is 25 m per minute at 30 s dipping time. An automated circulation and filtration units make sure of low liquor consumption during processing

(continued)

Table 1 (continued)

Innovation	Developed by	Principle of functioning
Magnoroll	J. Zimmer Maschinenbau, Austria	It is multipurpose coating machine with exchangeable modules. A variety of coating material like liquid, pastes, foams can be applied on a wide range of materials like textiles (carpets, non-woven), foils, glass, plastics, etc. Magnoroll can also provide pre-and-post coating of textiles in pigment digital printing processes to give higher rubbing fastness
MCD/3 Space Dyeing Machine	Superba Sas, France	It applies dye with a high pressure spray technique in one pile/one colour process. The machine is highly efficient with reduced water consumption, zero wastage of dyestuff, no colour migration and high colour fastness

a. **Composite Mills**: About 276 composite mills are currently operational in the country. They account for about 3% of the total output of the Indian textile sector. These mills are comparatively in large scale that have integrated spinning, weaving and fabric finishing departments.

b. **Spinning**: It is the most technologically efficient sector of Indian textile industry. However, the average unit sizes are relatively small and technologically outdated compared to other global competitors.

c. **Weaving and knitting**: These industries are engaged in converting the yarns into woven or knitted fabrics. These sectors are splintered into mostly small-scale and labour-intensive units. This sector comprises approximately 380, 000 power-loom enterprises that are small firms with 4–5 looms as its average capacity, 3.9 million small-scale handloom enterprises, with about 1.7 million looms that operate and just 1,37,000 looms in various composite mills. Shuttle-less looms account for less than 1% of the total production capacity.

d. **Wet Processing**: Wet processing includes preparation, dyeing, printing and finishing units that process textile material from fibre to garment stage. Out of nearly 2,300 processing units operating in India, 2,100 are independent and 200 are integrated units with spinning, weaving, knitting and Processing.

6.1   Reasons—Why Sustainability is not Main Stream in India?

Since these units are mostly small-scale labour-intensive units, sustainable business practices and processes are not applied. The reasons which prevent Indian Textile industry from embracing sustainable processing methods are

a. **Small-scale industry**: Most of the producers in the textile industry are small-scale unorganised firms which cater to the demands of big apparel exporters. Textile wet processing is both labour and chemical intensive. Most of these small-scale firms are located in the semi-urban regions or the outskirts of cities [63] where they easily miss from the notice of regulatory authorities. The effluents released by these units are mostly unchecked and untreated causing severe environmental harm to the vicinity.

b. **Government regulation**: India's stringent policies and regulations are applicable to large-scale firms and mills. However, being small-scale operating units, these firms fall under the 'unorganised' sector and are often able to evade such strict regulations and policies. Even though there have been efforts to implement regulations and encourage compliance from the small-scale textile sector, there is a long journey ahead [64, 65]. Existing policies are not as effective due to one rule for all conditions. Sometimes the paperwork and bureaucratic red tapes are too time-consuming and cost intensive for the small-scale firms to undertake at regular intervals.

c. **Market competition**: The large number of firms in the market makes it a very competitive environment. All the firms are competing for the business to generate more revenue and profit. The labour-intensive industry can counter any mechanical requirement by using more human resources to meet the demand of the buyer in lesser time and lesser price than the competitors to ensure the business keeps coming to them.

d. **Demand from buyers**: Buyers/bigger apparel firms provide business to the manufacturers who can provide the most competitive prices in lesser lead time. Sustainability is not the highest concern to meet the buyer demands in the least possible time, while making profit. Infrastructure bottlenecks in India lead to higher lead-times in the supply chain. Sustainable machinery is a costly investment. The small-scale businesses do not have much resources to spend on them.

e. **Organisational barriers**: Adoption of sustainable business practises are also influenced by the existing technologies available in the firm. The mentality: if something is not broken, don't fix it is followed. Most technological advancement is seen in the weaving sector as it increases efficiency and profit. However, the fabric finishing sectors depend on chemical processing to get results in short time. Additionally, these firms employ workers on a daily wage basis. In India, human resource is easily available. Workers do not always support introduction of smart AI-enabled automation technologies which may reduce labour force [66].

f. **Economic barriers**: Most of these small-scale firms make very less profit on the business they do. They do not generate much profit to invest in the innovative state-of-the-art technologies available in the market. Rising prices of raw materials, dye stuff, etc. have resulted in more economic strains for the textiles industry [67]. As we have seen previously, most of the advanced machines are manufactured in European countries like Germany, Austria, France, etc. There is a huge initial expense involved in purchasing these machineries, importing it to India, setting it up and subsequent maintenance [68]. Most of the small-scale firms are not able to make such purchases and investments.

g. **Lack of awareness**: Most of these small-scale firms lack the awareness of environmental degradation and effect of the subsequent processes practices by them on the environment. The daily wage workers are uneducated or very less educated; hence they are not aware or concerned about environmental impact.

6.2 Suggestions to Make Sustainability Mainstream in Indian Textile Industries

Being one of the largest exporters of dyes, pigments, textiles and apparel, Indian manufacturers have to opt for sustainable manufacturing processes—the sooner the better. Some suggestions to make sustainability mainstream in Indian textile industry:

a. More vigilant and customised government regulations and policies for the small-scale firms with strict financial consequences for breaking them. One fit for all policies are not as easy for the small-scale units to undertake. Smaller, simpler versions of the existing policies may encourage the small-scale firms to take the environmental regulations seriously.

b. With the rise in price of raw materials, buyers have to be willing to pay more for the fabrics developed and finished using sustainable products. They may market products made from these fabrics as high-end fashion lines which will make them more desirable to the end consumer and hence increase the demand of these fabrics.

c. The firms have to be open to experiment more sustainable options for long-term planning. The existing technologies working effectively does not mean they cannot look into other technologies which are more environmentally efficient. The firms also have to come to an understanding with the trade unions regarding the livelihood of the existing workers. These workers can be re-skilled and utilised within the existing worker pool instead of being fired. They may be re-skilled in operating the new technologies.

d. Economically, these small-scale firms would need more financial incentives from the government and easily available loans for purchasing the current sustainable machineries. There may be some exemption from taxes and other such benefits for the firm on purchase and use of these sustainable technologies for production.

e.   The firms need to be aware of their production practices and related impacts on environment. They have to follow government regulations for waste management strictly. Additionally, they should educate the workers on the benefits of sustainable production procedures for their health and the environment.

7   Summary and Conclusion

Different aspects of sustainability under each and every step of textile production can be summarised as given below.

- Factors of production including consumption of water and energy
- Availability and accessibility to sustainable raw materials
- Impact of industrial waste generation
- The corporate social responsibility towards the society at large
- Mainstream chemicals used like various dyes and coatings
- Occupational health hazards of employees of company and its consumers
- Animal welfare and ethical concerns, e.g. sourcing of wool, down and leather.

The main concerns of textile and apparel industries while producing sustainable products include child labour, occupational safety and fair wages in particular. Countertrends to fast fashion like eco fashion, green fashion, and fair trade labels, materials of high quality with durability, eco-textiles and recyclable textiles and packaging are becoming prominent in the fashion industry due to the increasing awareness of consumers. The feasible solutions for circular economy are recycling of discarded textiles and waste products. Designing of products utilising AR and CAD/CAM is also paramount. Most of the apparel industries are massively reducing waste in the process of their production by following digital sampling method, automated pattern grading and cutting, print placement and print on demand business models for customised manufacturing.

Sustainability has now become a buzzword across all manufacturing industries. Sustainability is the call of the hour in view of the current impact of fashion industry on the environment. Sustainability and fashion are both the driving forces for these revolutionary innovations to be implemented, which contribute to greener textile factories and goods that are more environmentally friendly. Today's consumers are well aware of the products which are ecologically and socially acceptable and consequently seek sustainable solutions in both textile as well as apparel industries. Under this circumstance, not only tapping of innovation potential but also promotion of sustainable development is the key commitment of Manufacturers today. The strategies for sustainability are ranging from sustainable process design to product designs, using innovative materials that offer reduce, reuse and recycling. However, the social and environmental challenges in the global textile value chain can be effectively solved if politics, businesses, global society and consumer demands pull together towards a common goal.

References

1. UN Documents (1987) Our common future: report of the World Commission on Environment and Development. http://www.un-documents.net/ocf-02.htm. Accessed 27 June 2021
2. Michael B-E (2007) Defining sustainability by the green imperative (244). https://www.resurgence.org/magazine/article85-defining-sustainability.html. Accessed 27 June 2021
3. Sen M (2020) How the textile industry pollution is affecting the environment. https://www.fastnewsfeed.com/health/how-the-textile-industry-pollution-in-affecting-the-environment/. Accessed 19 June 2021
4. Madhav S (2021) The dark side of colourful textiles. https://www.downtoearth.org.in/blog/water/the-dark-side-of-colourful-textiles-76505. Accessed 10 June 2021
5. Adhikary S (2021) Break for sustainable fashion in India on the horizon as pollution and wastage concerns rise. https://www.newindianexpress.com/magazine/2021/jan/03/break-for-sustainable-fashion-in-india-on-the-horizon-as-pollution-wastage-concerns-rise-2243759.html. Accessed 17 June 2021
6. Wicker A (2021) Fashion is not the 2nd most polluting industry after Oil. But what is it? https://ecocult.com/now-know-fashion-5th-polluting-industry-equal-livestock/#:~:text=Ok%2C%20so%20we've%20decided,the%20fashion%20industry%20more%20polluting. Accessed 12 June 2021
7. Anonymous (2019) How much do our wardrobes cost to the environment? https://www.worldbank.org/en/news/feature/2019/09/23/costo-moda-medio-ambiente. Accessed 23 May 2021
8. Bhardwaj T (2021) World environment day 2021—Challenges in managing toxic wastewater from textile industry and solutions. https://www.financialexpress.com/lifestyle/science/world-environment-day-2021-challenges-in-managing-toxic-wastewater-from-textile-industry-and-solutions/2265265/. Accessed 25 June 2021
9. Brown R (2021) The environment crisis caused by textile waste. https://www.roadrunnerwm.com/blog/textile-waste-environmental-crisis. Accessed 10 May 2021
10. Grant M (2020) Sustainability. https://www.investopedia.com/terms/s/sustainability.asp. Accessed 25 June 2021
11. Anonymous (2014) Organic cotton a sustainable choice. https://www.fibre2fashion.com/industry-article/7251/organic-cotton-a-sustainable-choice. Accessed 10 May 2021
12. Anonymous. Organic Cotton. http://aboutorganiccotton.org/. Accessed 11 May 2021
13. Anonymous (2020) Is organic cotton sustainable? Here's what you need to know. https://www.sustainme.in/blogs/news/is-organic-cotton-sustainable-here-s-what-you-need-to-know. Accessed 11 May 2021
14. Hadjiosif S (2019) Is organic cotton sustainable? Pros and cons. https://www.terramovement.com/is-organic-cotton-sustainable/ . Accessed 25 May 2021
15. Saxena S, Raja ASM, Arputharaj A (2017) Challenges in sustainable wet processing of textiles. In: Muthu S (ed) Textiles and clothing sustainability, textiles science and clothing technology. Springer, Singapore. https://doi.org/10.1007/978-981-10-2185-5_2. Accessed 30 May 2021
16. Whewell C (2016) Textile—Dyeing and printing. https://www.britannica.com/topic/textile/Dyeing-and-printing. Accessed 17 May 2021
17. Jaganmohan M (2021) India: dyes and pigments production volume 2021 | Statista. https://www.statista.com/statistics/726947/india-dyes-and-pigments-production-volume/. Accessed 19 May 2021
18. Anonymous (2021) Annual report 2020–21. https://chemicals.nic.in/sites/default/files/Annual_Report_2021.pdf. Accessed 30 May 2021
19. Varadarajan G, Venkatachalam P (2016) Sustainable textile dyeing processes. Environ Chem Lett 14:113–122. https://doi.org/10.1007/s10311-015-0533-3.Accessed10May2021
20. Tanapongpipat A, Khamman C, Pruksathorm K, Hunsom M (2008) Process modification in the scouring process of textile industry. J Clean Prod 16(1):152–158. ISSN 0959-6526. https://doi.org/10.1016/j.jclepro.2006.06.016

21. Costa SA, Tzanov T, Carneiro F et al (2002) Recycling of textile bleaching effluents for dyeing using immobilized catalase. Biotech Lett 24:173–176. https://doi.org/10.1023/A:1014136703369

22. Eren HA et al (2009) Enzymatic one-bath Desizing—Bleaching—Dyeing process for cotton fabrics. Text Res J 79(12):1091–1098. https://doi.org/10.1177/0040517508099388

23. ColorZen News (2018) ColorZen wins innovative competitions, innovation in textiles, Copenhagen. https://www.innovationintextiles.com/colorzen-wins-innovation-competition/

24. Zhang S, Ma W, Ju B, Dang N, Zhang M, Wu S, Yang J (2006) Continuous dyeing of cationised cotton with reactive dyes. https://doi.org/10.1111/j.1478-4408.2005.tb00270.x

25. Prabu HG, Sundrarajan M (2006) Effect of the bio-salt tri-sodium citrate in the dyeing of cotton. https://doi.org/10.1111/j.1478-4408.2002.tb00370.x

26. Warren L (2021) Scientists use bacteria to make sustainable Indigo Dye. Sour J. https://sourcingjournal.com/denim/denim-innovations/sustainable-indigo-dye-bacteria-korea-advance-institute-science-technology-285790/. Accessed 25 June 2021

27. Guest Editorial SJ (2021) Advance Denim introduces bioblue Indigo to safely enhance eco-friendly dyeing. Sour J. https://sourcingjournal.com/denim/denim-innovations/advance-denim-bioblue-indigo-dyeing-sustainable-denim-safety-reduction-275626/

28. Shin Y, Choi M, Yoo DI (2014) Eco-friendly indigo reduction using bokbunja (Rubus coreanus Miq.) sludge. Fash Text 1(6). https://doi.org/10.1186/s40691-014-0006-5

29. Saikhao L, Setthayanond J, Karpkird T, Bechtold T, Suwanruji P (2018) Green reducing agents for indigo dyeing on cotton fabrics. J Clean Prod 197:106–113

30. Saling P et al (2002) DyStar textilfarben GmbH, eco-efficiency analysis by BASF: the method. Int J Life Cycle Assess 7:203–218

31. Bhosale N, Jadhav B (2014) Sustainability in Indigo Dyeing. Texnote. http://texnote.blogspot.com/2014/06/sustainability-in-indigo-dyeing.html

32. American Association of Textile Chemists and Colorists (1953) The application of vat dyes. In: AATCC monograph, no 2, p 230

33. Hebei Rising Chemical Co., LTD, Shijizhuang, Hebei, China

34. Srihari Balakrishnan, Managing Director, KG Denim Ltd., Coimbatore, India, Report on Sustainable Denim Processing

35. Parker P (2012) Director of engineering, morrison textile machinery, Fort Lawn, SC, A report on Denim Processing

36. Wambuguh D, Chianelli R (2008) Indigo dye waste recovery from blue denim textile effluent: a by-product synergy approach. New J Chem 32:2189–2194

37. Warren L (2021) Scientists use bacteria to make sustainable Indigo Dye. Sour J. https://sourcingjournal.com/denim/denim-innovations/sustainable-indigo-dye-bacteria-korea-advance-institute-science-technology-285790/

38. Kan CW (2015) Plasma treatments for sustainable textile processing. In: Sustainable apparel. Woodhead Publishing Series in Textiles, Pp 49–118. ISBN 9781782423393. https://doi.org/10.1016/B978-1-78242-339-3.00003-0

39. Kavyashree M (2020) Printing of textiles using natural dyes: a global sustainable approach. https://doi.org/10.5772/intechopen.93161. https://www.intechopen.com/chapters/72788

40. McKeegan D (2019) Sustainability is the paradigm of digital textile printing. https://www.fespa.com/en/news-media/features/sustainability-is-the-paradigm-of-digital-text

41. Knochel R (2020) Sustainable approaches in textile printing. Planet Adv. https://www.planet-advertising.com/en/2020/05/14/sustainable-approaches-in-textile-printing/

42. Chemical & Engineering News. ISSN 0009-2347. Copyright © 2021 American Chemical Society

43. Anonymous. Textile finishing processes. https://www.britannica.com/topic/textile/Textile-finishing-processes. Accessed 10 June 2021

44. Anonymous (2019) The environment impacts of textile finishing. https://www.evolvedbynature.com/the-environmental-impacts-of-textile-finishing. Accessed 12 June 2021

45. Körlü A (2018) Use of ozone in the textile industry, p 2. https://www.intechopen.com/books/textile-industry-and-environment/use-of-ozone-in-the-textile-industry. Accessed 18 May 2021

46. Bishop M (2014) Ozone finishing for denim reduces environmental impact, processing costs and processing time. https://risnews.com/ozone-finishing-denim-reduces-environmental-imp act-processing-costs-and-processing-time. Accessed 22 May 2021
47. Bahtiyari MI, Benli H (2016) Ozone bleaching of cotton fabrics with the aid of ultrasonic humidifier. Cellulose 23:2715–2725. https://doi.org/10.1007/s10570-016-0978-y
48. Yiğit İ, Eren S, Eren HA, Ozone utilisation for discharge printing of reactive dyed cotton. Color Technolgy. https://doi.org/10.1111/cote.12306
49. Radhia JA (2016) Plasma technology—A sustainable approach to textile processing, pp 2–3. https://www.researchgate.net/publication/321254953_Low-temperature_Plasma_Techno logy_A_Sustainable_Approach_to_Textile_Processing. Accessed 20 June 2021
50. Parmar S, Malik T (2007) Use of plasma technology in textiles. https://www.fibre2fashion. com/industry-article/1798/use-of-plasma-technology-in-textiles#:~:text=Plasma%20treatme nts%20have%20been%20used,conventional%20fabrics%20to%20advanced%20composites. Accessed 17 May 2021
51. Anonymous (2013) Surface treatment finds eco-friendly solution in plasma. https://www.fibre2 fashion.com/industry-article/7035/surface-treatment-finds-eco-friendly-solution-in-plasma. Accessed 25 May 2021
52. Eren H, Avinc O, Eren S (2017) Shield square captcha. https://iopscience.iop.org/article/ https://doi.org/10.1088/1757899X/254/8/082011#:~:text=Supercritical%20carbon%20diox ide%20(scCO2)%20waterless,supercritical%20carbon%20dioxide%20(scCO2). Accessed 18 May 2021
53. Tiwari S (2010) Dyeing of PET fibre using supercritical carbon dioxide. https://www.fib re2fashion.com/industry-article/4932/dyeing-of-pet-fibre-using-supercritical-carbon-dioxide. Accessed 21 May 2021
54. Stepankova M, Wiener J, Dembicky J (2010) Impact of laser thermal stress on cotton fabric. Fibers Text East Eur 18:70–73
55. Lamar D (2020) Laser finishing for textile materials | Textile World. https://www.textilewo rld.com/textile-world/features/2020/07/laser-finishing-for-textilematerials/#:~:text=The%20l aser%20beam%20must%20be,beam%20onto%20the%20material%20surface. Accessed 19 May 2021
56. Hung O-N, Kan C-W (2017) Effect of CO_2 laser treatment on the fabric hand of cotton and cotton/polyester blended fabric. Polymers 9(11):609. https://doi.org/10.3390/polym9110609
57. Prince P, A, P, R (2007) Nano-finishing of textiles (TT-03). https://www.fibre2fashion.com/ industry-article/1505/nano-finishing-of-textiles#:~:text=The%20nano%20technology%20f inish%20creates,the%20life%20of%20the%20garmen. Accessed 19 May 2021
58. Periasamy A, Militky J, Sachidhanandham A, Duraisamy G (2021) Nano technology in textile finishing—Recent developments, pp 4–6. https://www.researchgate.net/publication/349109 717_Nanotechnology_in_Textile_Finishing_Recent_Developments. Accessed 29 June 2021
59. Singh M, Dua JS, Menra M, Soni M, Prasad DN (2016) Microencapsulation and its various aspects: a review. Int J Adv Res 4:2094–2108. https://doi.org/ 10.21474/IJAR01/726. https://www.researchgate.net/publication/305336471_MICROENCA PSULATION_AND_ITS_VARIOUS_ASPECTS_A_REVIEW. Accessed 19 May 2021
60. Saleem H, Zaidi J, Syed (2020) Sustainable use of nanomaterials in textiles and their envi- ronmental impact. https://www.researchgate.net/publication/347770233_Sustainable_Use_ of_Nanomaterials_in_Textiles_and_Their_Environmental_Impact. Accessed 24 June 2021
61. Anonymous (2019) About ITMA. https://itma.com/about-itma. Accessed 19 May 2021
62. Dhanabhakyam M (2007) Indian textile industry—An overview—Free industry articles— Fibre2fashion.com. https://www.fibre2fashion.com/industry-article/2363/indian-textile-ind ustry-an-overview. Accessed 19 May 2021
63. Anonymous (2020) Contemporary issue in textile industry. https://textilevaluechain.in/ in-depth-analysis/articles/textile-articles/contemporary-issues-in-textile-industry/. Accessed 9 May 2021
64. World bank (2020) Environmental and social systems assessment (ESSA). https://msme.gov. in/sites/default/files/EnvironmentalSocialAssessmentRAMP.pdf. Accessed 30 June 2021

65. Ghatak A (2019) Environment regulations and compliance in the textile dyes Sector of Gujrat: case of Ahmedabad Cluster. https://www.researchgate.net/publication/335772000_Environme ntal_Regulations_and_Compliance_in_the_Textile_Dyes_Sector_of_Gujarat_Case_of_Ahm edabad_Cluster. Accessed 30 June 2021
66. Zapfl D (2019) How digitisation affects the textile industry. https://www.lead-innovation.com/ english-blog/how-digitisation-affects-the-textile-industry. Accessed 10 May 2021
67. Narashimhan TE (2021) Textile exporters asks buyers to increase prices as raw material costs rise. https://www.business-standard.com/article/economy-policy/textile-exporters-asks-buyers-to-increase-prices-as-raw-material-costs-rise-121032300442. Accessed 9 May 2021
68. Dave N (2019) Increasing textile machinery cost a hurdle for newcomers. https://www.fin ancialexpress.com/industry/increasing-textile-machinery-cost-a-hurdle-for-newcomers/145 5317/. Accessed 10 May 2021

Study on Product Safety of Children's Apparel

D. Sanjeevana and R. Thenmozhi

Abstract The safety of children's clothing is a primary concern. Different types of trims and attachments are being used on children's garments to make them more appealing and appropriate. A safe product is one that poses no or minimal risk, as well as reasonably predictable product use and the requirement to maintain a high level of consumer protection. Organic cotton products for children are safer and more environmentally friendly. According to this survey, the use of eco-friendly sustainable textiles, dyeing, printing, trims and accessories on children's products is growing. This research also reveals difficulties with currently available products, as well as the opportunity for new product creation and long-term sustainability.

Keywords Product safety · Organic cotton children's garment · Sustainable dyeing · Printing · Interlock fabric · Trims and accessories

1 Introduction

Clothing safety for children has always been a top priority. Fabric selections, fasteners, fit and ease, and the trimmings used in children's clothes are all essential factors when creating children's apparel. The changing shape of a growing child, as well as the fluctuating proportions of various body parts, are further factors to consider for a designer working on children's clothing.

A possible source of injury is classified as a hazard. The most common risks in children's clothing have been identified. An overview of the nature of the hazard and how it poses a risk to children is provided for each. The main hazards are choking and swallowing, sharp edges and points.

Laboratory testing is required to ensure that the product complies with all applicable regulations. In other cases, laboratory testing is necessary to ensure the item's safety, quality and functionality, as well as the item's subsequent consumer acceptance.

D. Sanjeevana · R. Thenmozhi (✉)
Department of Apparel & Fashion Design, PSG College of Technology, Coimbatore 641004, India
e-mail: rtm.afd@psgtech.ac.in

© The Author(s), under exclusive license to Springer Nature Singapore Pte Ltd. 2022   157
S. S. Muthu (ed.), *Sustainable Approaches in Textiles and Fashion*,
Sustainable Textiles: Production, Processing, Manufacturing & Chemistry,
https://doi.org/10.1007/978-981-19-0538-4_7

Children under the age of three are the most sensitive to trims and little bits that can easily detach from garments, since they are still mouthing objects and lack the ability to form judgements. Testing with calibrated equipment and to an industry-based standard is therefore highly advised to ensure that the small components or trims will remain attached for the duration of the garment's useful life.

2   Litreature Review

Haque et al. studied **quality assessment of children's clothing**, and they report their study, the limitations regarding the quality of kid's apparel towards parents in the present condition of Bangladesh, and also that the study creates awareness among parents related to apparel quality on children's health.

Kothari et al. [1] studied the communication of kids' wear safety to parents. According to this study, the majority of Indian children's clothing retailers believe it is critical to explain children's clothing safety to parents, and that government promotions are the best way to do so. Indian retailers believe that a safety mark for children's clothing, such as ISI or Eco Mark, should be implemented so that parents can readily distinguish between safe and harmful children's clothing.

Farhanin et al. [2] proposed Children's product safety in Malaysia which reported that it is the responsibility of every person at every level to prevent the health of children. The distributor also has the responsibility of having the clear status of the product before it is distributed to any other place.

Hasan et al. discussed product safety requirements for children's clothing. Their study reveals that children's clothing must pass basic safety rules before being sold, and also, they mentioned in their result that the people are unaware of this, and they discussed that the branded products are committed to providing quality products.

Das [3] examined the study's quality difficulties in South African garment merchandising. Their research demonstrates the potential usefulness of using FOM in each segment and individual product line, as well as the necessity for a full study in each of the various clothing segments, such as menswear, ladies' wear, children's wear and so on.

The effect of fabrics and designs on the physical comfort of children's clothes in the Accra Metropolis was studied [4]. In this study, they found that knowledge of parents and fashion designers on the properties of fibers and safety measures in apparel manufacturing are crucial in safeguarding the comfort ability of children's apparel.

Georgios et al. [5] studied the screen printing of cotton with natural pigments and reported that the important properties of the printing paste, such as rheological and physical properties, pH, conductivity and viscosity, were measured for natural pigments.

Ashkumar et al. 'Application of natural dyes on textiles'. In this study, pre-mordanted or simultaneous mordanted, the colour depends not only on natural colourant but also on the mordant and mordanting assistance. It was also studied that the specific pre-mordanting methods, colour matching databases for natural dyeing of any particular textile material and match prediction are possible with a computerised programme.

2.1 Requirements for Children's Clothing

Children's clothes are made with safety and comfort in mind, with fabric and design for the physical comfort of babies. The sensitive skin of children reacts to some materials, making them uncomfortable to wear. Action should be encouraged, and children's movements should be unrestricted. Clothing should give a child a sense of security and comfort. Discomfort dresses are annoying and make someone stand out in a crowd. Relaxation, self-help characteristics, attractiveness and growing aspects should all be considered while picking a design for children's clothing. Children are grouped into age groups, each with its own set of requirements.

2.2 Analysis of Children's Clothing

Even though some publications have suggested that children's clothing might cause damage, medical data on garment-related injuries in children is limited. As a result, the goal of this research was to compile a list of the descriptive features of child clothing-related injury cases. Children aged 3–6 were most vulnerable to clothing-related injuries. Clothing-related injuries were found in around 51% of children's hospital cases. The most likely place for the incident to occur is at home. Furthermore, zipper injuries are the most common, with 24 documented deaths in Eastern Europe owing to clothing injuries.

2.3 Garment Safety for Children's Clothing

Choking, asphyxiation and swallowing are all more likely in children aged 36 months and under when they come into contact with small items. Snaps, studs, rivets, buttons, zipper components, appliques, sliders, dungaree clasps (Hasps), rosettes, bows, belt fastenings, toggles, small textile components, decorative labels on the exterior of the garment (tabs) and rouleau loops are all examples of fastenings that can come loose on children's clothing.

2.4 Safety Measures

To ensure that no stray metal contaminates the product all that is required is preventative measures. Detection is performed as a safety check to make sure the product is safe. Compliance with metal contamination policies necessitates management's unwavering commitment as well as operative instruction. Factory managers should start by determining the risks of metal contamination posed by their operations.

2.5 Fabric and Comfort

There are two of the most crucial aspects to consider while selecting children's apparel. Some consumers have a taste for certain fabrics, while others do not, but everyone wants to be comfortable. But there's one more thing that needs to be justified. It is necessary to use fabric that is suitable for babies and infants. Young people, on the other hand, may choose their outfits depending on the fabric, colour, and pattern.

2.6 Sustainable Dyeing

Two natural dyes (alkanet and rhubarb) on wool, silk, cotton and flax. The dye concentration, thickening agent nature, fixation type, mordant concentration and mordant type were all investigated. The K/S value of the printed items, as well as their overall fastness characteristics, were evaluated. The thickener utilised as Meypro gum has the highest K/S value, according to the data. Natural colour powder is gradually added to the printing paste from 10 to 40 g/kg.

To examine how mordants affected colour development, alkanet dye was utilised. Printing paste concentrations of up to 20 g/kg yielded satisfactory results. The colourfastness findings varied depending on the mordants used, and they ranged from excellent to very good.

2.7 Sustainable Printing

Natural dyes are becoming more popular in eco-textiles and fashion as people become more aware of the need to apply sustainability to the environment. As a result of the above challenge, proof of algal biomass, a unique, environmentally friendly, biodegradable, non-carcinogenic and long-lasting colourant, has been discovered. Fresh algal biomass is used to extract red pigment and phycoerythrin, which is then used in a pigment-printing method that combines natural and synthetic printing.

2.8 Natural Dye Printing on Textiles: A Global Sustainable Approach

The fashion industry's approach to sustainability has significantly evolved as a result of globalisation. Synthetic dyes and fabrics have been the focus of fast-changing fashion trends, but these items have failed to bridge the gap between sustainability and environmentally responsible design.

As a result, the focus of this chapter is on creating a sustainable, eco-friendly screen printing technology that utilises natural colours on silk and cotton materials.

This chapter is focused on sustainability design, which allows for a lot of natural colours to be used in everyday fashion. This research focuses on the colourfastness of the dyes utilised.

2.9 Supply Chain Strategies and the Effects of the COVID-19 Epidemic

According to the findings, the strategies of manufacturing flexibility, expanding the source of supply and creating backup suppliers all have a substantial positive influence on minimising the COVID-19 pandemic's implications in the RMG supply chain.

It aids industry managers in recovering from supply chain disruptions by recognising and defining the consequences and procedures needed to resolve the substantial supply network disruptions caused by the COVID-19 pandemic. This paper is one of the few early attempts to examine the ramifications of the COVID-19 epidemic and approaches to deal with the impacts in the supply chain environment as a theoretical contribution.

3 Materials and Method

3.1 Material

Body fabric—Organic interlock.

Sewing thread—Poly core spun thread.

Accessories—mock wood buttons and snap buttons.

Printing—Natural dye powder, eco-friendly printing.

3.2 Organic Interlock

Interlock is soft to the touch and pleasing to the eye. This fabric has an incredibly utilitarian nature. Children's clothes, particularly for new borns, are also priced for cotton interlock.

Organic interlock is an ideal knit fabric. It's exceptionally soft, warm and flexible, making it ideal for infant clothes. It's warmer than rib knit or cotton jersey. It's made of two knits, which makes it thicker and heavier than a single knit. Interlock knit is slightly thicker and more stable than other knits, and it doesn't curl at the edges, making cutting and sewing very easy.

3.3 Double Knit

Double knit construction is done mainly to avoid the intense burning flammability. According to the US CPSc (Consumer Product Safety Commission), children who sleep in loose-fitting and over-sized cotton clothing are in danger. Because children's clothing can easily catch fire. The danger with loose-fitting clothing is that there is an air pocket between the clothing and the child's skin, which helps the fire burn.

3.4 Specifications for Button

Mock Wood

- It is sustainable and eco-friendly
- It is biodegradable
- It is water absorbing expansion.

Snap Button

- Snaps are more durable to use than Velcro or buttons
- When the baby lies on a snap, it's not something that leaves a big mark on the skin. Buttons and zippers are thicker.
- When you undress the baby, you can just pull the one side apart.

3.5 Printing Material

Pigments are solid powders of colour that are either organic or inorganic. They are insoluble colourants, primarily of mineral origin, that are used to colour metals, wood, stone and textiles. These pigments do not bind to fibres in any way. They're utilised for printing with the binder system. By producing a layer on the fabric surface, the binder guarantees that pigment particles stay on the fibre.

On heat fixation, long macro-molecules of binder, along with binder and thickening, create three-dimensional linkages. The printed fabrics are heat-cured at 150 °C during the fixing process. This can be done in a heat press or hot mangle, or in a curing oven in an industrial process.

Natural dyes have high value when it comes to dyeing garments. Because they are natural and eco-friendly, they cause no harm to the environment, and the effluents produced after their application to textiles for colouring cause less pollution.

Grapes are utilised in both residential and industrial settings to make soft drinks and other consumer products. After extracting the juice from the grape, however, the solid remnants are ignored and discarded. In this study, the skin of the grape produces lovely colours on protein materials like silk and wool. So, for an indigo colour, they used grape skin powder to dye the body cloth.

3.6 Garment Design

See Fig. 1.

3.7 Mock Wood Button

See Fig. 2.

4 Methodology

Quality test conducted for kid's apparel

- Fibre analysis.
- Review of labelling/fibre composition marking.
- Colour fastness to laundering.
- Colour fastness to saliva.
- PH value.
- Free formaldehyde.
- Appearance after 3rd wash.
- Colour fastness to rubbing.
- Safety on kid's clothing-cord and drawstrings.
- Flammability.

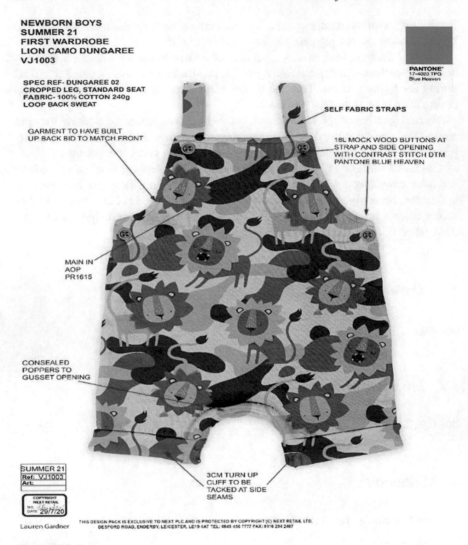

Fig. 1 Kids Romper

4.1 Fibre Analysis

Fibre analysis and testing is a complex discipline that requires a focused approach
to sample preparation in order to conduct a full analysis, with the results being
appropriately examined and applied correctly. Only an experienced approach will be
able to appropriately analyse the results.

Fig. 2 Mock wood button

4.2 Review of Labelling/Fibre Composition Marking

A label should have the following information: the producer's name, corporate name or designation, as well as the residence of the maker, trader or importer. The national manufacturer's industrial registration number must be included on products created in other countries. Traders can use registered trademarks on their products, but they must provide the earlier specified evidence. The tag will be made of durable material and will be sewn or adhered to the clothing permanently, by a similar lifecycle. If the material comes in a package, the labelling must be observable from the outside or appear on the package itself. Specifiable indicators such as protection symbols, non-shrink, fireproof, and impermeable must be clearly separated by product.

4.3 Colour Fastness to Laundering

The dyeing fastness of textile materials for garments and home textile products follows one or more simulations of household and commercial washing. GB/T3921-2008, ISO 105 C10:2006 and AATCC 61:2010 are typical test standards. From a practical standpoint, it is important to compare the three standards.

4.4 Saliva Colour Fastness

Determine the colour of the textile's resilience to the effects of saliva. This test is only required for fabrics used in children's and infants' clothing.

4.5 PH Value

The pH value is used to determine how much acid and alkali are still present in clothing textiles. Apparel textiles that come into direct contact with the skin have a higher pH value.

4.6 Free Formaldehyde

This test method is used to determine the amount of free formaldehyde (HCHO) present in emulsion polymers without disrupting the formaldehyde equilibrium. Acrylonitrile butadiene, carboxylate styrene-butadiene and polyvinyl acetate emulsion polymer were used to test the technique. This test procedure may also be relevant to other types of emulsion polymers. This test method's established operating range is 0.05–15 ppm formaldehyde.

4.7 Appearance After 3rd Wash

This test is used to evaluate the appearance of fabrics or garments after they have been washed three times. The specimen's appearance is rated in comparison to other reference specimens under standard lighting and viewing conditions.

4.8 Rubbing Fastness

It refers to the resistance of the colour of textile to abrasion or staining to other textiles. As a rule, dark colours achieve lower rub fastness than light colour.

4.9 Safety of Kid's Clothing—Cords and Draw String on Kids Garment

Cords and drawstrings are used in different parts of garments, such as hoods and waists, for fastening the garments or as simple decorations. If not designed properly, drawstrings and cords on kids' clothing can present a safety hazard that potentially leads to injury or death if the cord or drawstring catches on fixed or moving objects.

4.9.1 Flammability

Normal Flammability (Class 1).

Clothing without nap, pile, tufting, flock, or any other sort of raised fibre surface has a flame spread duration of 3.5 s or greater. More than 7 s for children's clothing with various sorts of raised fibre surfaces, assuming the flame intensity is low enough to avoid igniting fire.

Intermediate Flammability Class 2.

For children's clothing with any form of raised fibre surface, the flame spread period is 4–7 s.

Rapid and Intense Flammability (Class 3).

Clothing with or without nap, pile, tufting, flock, or other sort of raised fibre surface has a flame spread period of less than 3.5 s. This third type of fabric is extremely flammable.

4.9.2 Pull Test of Garment Accessories

One of the most common testing methods is the simple pull test. An inspector will often use a mobile pull gauge or a table top pull test machine to apply a certain force to an item for a set period of time to conduct this test. The result of the test is 'fail' if the accessory detaches from the garment.

4.9.3 Fatigue Test of Accessories

Many product inspections include lifetime testing, which involves using the product as intended for a number of cycles. Garments are no exception. During clothes inspection, importers frequently request a fatigue test to ensure the performance and longevity of accessories.

4.9.4   Stretch Test for Elastic Fabric and Straps

A stretch test is used to assess the stretch, elongation and recovery of elastic fabric or accessories on a garment, such as straps and ties. During the test, the inspector stretches the elastic by hand, looking for any breaks or stretches in the elastic fibres, as well as any loss of function after stretching.

4.9.5   Check Button Holes for Stitching Defects

Fatigue testing and pull tests can be used to check the button's function and durability. However, a button without a functional buttonhole is useless. Because there are no specific on-site tests for buttonhole stitching, the inspector should keep an eye out for a variety of quality defects related to buttonhole stitching during the garment inspection.

5   Discussion

Dungaree clothing is a one-piece outfit. A dungaree is a combination of a short- or long-sleeved shirt with shorts. The main intention of every parent is to find outfits that are comfortable and, primarily, safe for their kids. When it comes to comfort, opting for one-piece items is comfortable for a child as well as convenient for parents. This study shows the difference between commercial dungarees and developed dungarees.

The developed dungaree design is based on comfort, environment-friendly, safe for kids, no harmful accessories or more buttons are used like in other dungaree garments. In this study, the author used organic interlock to give more durability than regular. The organic interlock provides additional comfort for the baby, wicks away moisture, and ventilates cool air, reducing the likelihood of the baby sweating or being overheated. Its weaving will also keep you warm in the colder months. The same little openings that allow a breeze to circulate through the material will also aid in thermal insulation.

Natural dyes are not damaging to the environment because they are derived from natural sources, and the dungaree that is on the market has a greater number of metal buttons and zippers at the centre of the garment. Buttons and zippers are thicker and leave a big mark on the skin. There is a possibility that the children may choke on the buttons, and normal rompers have a single piece of fabric that catches fire quickly, hurting their back, neck or shoulders if the children's clothes don't fit properly. Whereas in this research design, the author used mock wood buttons and double knit fabric, which has less flammability and is easy to wear. When you undress the baby, one can just pull the one side apart.

From this study, the main objective is to give importance to the subject's being protected from the harmful environment. In considering the hidden dangers, various tests and reports have been taken to prevent damage, and providing a safety garment to kids is very important.

The use of toxic and synthetic colours like azo and benzamine has been a source of global concern in the fast fashion industry. Children, nature and mankind are all negatively affected by these dyes. With the results of our experiment and research, it is clear that screen printing with indigo on cotton interlock fabric produces a promising colour and can be regarded a safe alternative to dangerous synthetic dyes.

References

1. Kothari V, Mathews S (2015) Communication of kids' wear safety to parents: Indian retailers' viewpoint. J IMS Group 12(2)
2. Asuhaimi FA et al (2017) Legal issues relating to online transaction: special reference to children product safety in Malaysia
3. Das S (2011) Quality issues related to apparel merchandising in South Africa. Nelson Mandela Metropolitan University, Diss
4. Dogbey R et al (2015) The effect of fabrics and designs on the physical comfort of children clothes in the Accra Metropolis. Choice 30
5. Savvidis G et al (2017) Screen-printing of cotton with natural pigments: evaluation of color and fastness properties of the prints. J Nat Fibers 14(3): 326–334

Extraction and Optimization of Natural Dye from *Madhuca Latifolia* Plant Bark Used for Coloration of Eri Silk Yarn

R. Sujatha

Abstract Eri silk is one of the four varieties of silk produced in India. Recently eri culture was introduced in Andhra Pradesh due to the availability of host plants. Eri silk fabrics are suitable in the fields of apparel, furnishing and home textiles. Due to the market potential for eco-friendly fabrics, it became mandatory to develop eco processing for eri silk. Eri is a wild silk that cannot be reeled, due to the piercing of cocoons by the moths. Hence, it is used for the production of spun yarn only. The importance of eri silk cannot be undervalued as a fine textile fibre. It has got certain outstanding textile properties and is unique in many respects [7]. The bark of *Madhuca latifolia* source was selected for the study as this source was not optimized for dyeing purposes. Alum, stannous chloride, ferrous sulphate, tartaric acid and chitin were selected for the study [8]. A colour flex spectrophotometer was used to assess the colour strength of the liquids as well as dyed yarns. The colour fading and staining due to exposure to serviceable conditions such as sunlight, washing, crocking and perspiration were also assessed using a colour flex spectrophotometer. *Madhuca latifolia* dyed samples exhibited colours ranging from pinkish brown, light brown to dark brown colours.

Keywords Natural dyes · *Madhuca latifolia* · Mordant · Eri silk · Colour strength · Fastness

R. Sujatha (✉)
Department of Home Science, Sri Padmavati Mahila Visvavidyalayam, Tirupati, India

© The Author(s), under exclusive license to Springer Nature Singapore Pte Ltd. 2022   171
S. S. Muthu (ed.), *Sustainable Approaches in Textiles and Fashion*,
Sustainable Textiles: Production, Processing, Manufacturing & Chemistry,
https://doi.org/10.1007/978-981-19-0538-4_8

1 Introduction

Eri silk is one of the four varieties of silk, which is known for its durability is a regular winter wrapper for Assamese people. Eri silk act like a cotton fibre made from eri silk cocoon. Eri silk is unique in its characteristics. It shows lustrous nature like cotton and in case of warmth and bulkiness, eri silk looks like wool. Eri silk has a natural off-white colour, so it delivers beautiful appeal. Eri silk can be dyed in different shades. It is the only silk which is drawn without killing the pupae, and hence qualifies as most eco-friendly and 'ahimsa silk' among other varieties of silk. Eri silk contains certain medicinal properties and hence it charged better price than mulberry silk in the market. Nowadays, there is a very good scope for the development of eri culture. Karnataka, Andhra Pradesh, Tamilnadu, Gujarat, Uttar Pradesh, Rajasthan and Punjab states have shown much potentiality in eri silk production. In recent years, eri culture had been started in Andhra Pradesh due to the suitability of its climatic condition. The Eri silk production is being encouraged by the former Andhra Pradesh and present Telangana state, Devarakonda village in Nalgonda district, and Shadnagar village in Mahabubnagar district, where Tapioca is cultivated abundantly, which is a feedstuff for Eri silkworms. In Andhra Pradesh, Peapally village in Kurnool district also involved in the Eri silk production which is introduced and encouraged by the Government through incentives. Indeed, Eri culture can be taken up by rural women in a small space to increase their family income [2]. The predictions for "Eri silk" are good, but its supply in the market is not sufficient as far as India is concerned. It is also called as "poor man's" silk, Eri silk produces a very strong and supple fabric. The Eri silk fibre is a substitute for woollens and is also used as chaddars or wraps. Eri silkworms are comparatively cheap and easy to maintain. In India, Assam and Meghalaya states are producing the majority (95%) of the world's Eri silk.

In the present scenario, the increasing concentration on the utilization of natural dyes on natural fibres is giving a very significant part worldwide due to the eco-friendly consciousness [4, 9]. Colouring of eri silk with a natural dye source improves the fabric's aesthetic values [1]. This eri silk has very potential and provides wide scope for development of creating diversified products, which can play a important role in upgrading of the rural economy as well as its demand in the market [7].

The bark of *Madhuca latifolia* is from a renewable source and is abundantly available in the local areas of the Chittoor district. The natural dye colourant obtained from *Madhuca latifolia* may be considered a very good potential source of eco-friendly natural dye for textile materials.

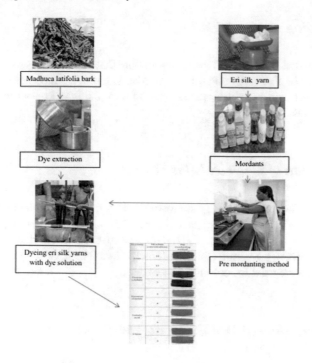

2 Selection of Materials

2.1 Selection of Yarn

Handspun eri silk was selected for dyeing with *Madhuca latifolia* bark.

2.2 Selection of Madhuca Latifolia Dye Source

The Indian butter tree (local name Mahua; *Madhuca latifolia* or *Madhuca langifolia*) is known for its multifarious utility. The timber and edible oil that comes out of the tree was valued. The tree was also found to be unique with varied medicinal properties and beneficial in minimise the skin allergies. *Madhuca latifolia* is an economically important plant of the family Sapotaceae. It played an important role in the tribal economy of Madhya Pradesh. It was also described as a pollutant tolerant tree. The bark of *Madhuca latifolia* was found abundantly in the local areas of Andhra Pradesh. So it was selected as a natural dye source for this study.

2.3   Selection of Mordents

Mordants form the link between dyestuff and fibre that allows the dye with no affinity for the fibre to be fixed. Among all the mordants, alum, stannous chloride, ferrous sulphate, tartaric acid and chitin were selected for the study as they were eco-friendly and did not cause any pollution.

2.4   Selection of Chemicals for Testing

The following chemicals of LR grade were used for conducting this research. Alum $(Alk(So_4)_2)$, Stannous Chloride $(SnCl_2)$, Ferrous sulphate $(FeSo_4)$, Tartaric acid $((CHOH–COOH)_2)$ and Chitin $(C_8H_{13}No_5)$ were used as mordants. Sodium Bi Carbonate $(NaHCo_3)$, Sodium Chloride $(Nacl)$, Urea (NH_2CONH_2) and Acetic acid (CH_3COOH) were used for colour fastness tests.

2.5   Optimization of Dye Extraction Methods

Aqueous, alkaline and acidic methods were experimented for dye extraction. In the aqueous method, water was the medium for the extraction of dye. The raw material was boiled in plain water and the dye was extracted. Water was made acidic with the addition of 0.1–1% of acetic acid by acidic method. The raw material was boiled in this medium. Optimization of alkali for each source series of experiments was conducted to optimize the alkali from 0.1 to 1%. The raw material was boiled in an alkaline medium. The solution was made alkaline with the addition of one percent sodium carbonate to water. The alkaline method was found to be the best method for the extraction of dye from the bark of *Madhuca latifolia*.

2.6   Optical Density

The dyeing performance of various dyed samples such as depth of shade, tone, etc. were measured with a computer colour flex spectrophotometer.

2.7   Selection of Suitable Wavelength

Absorbance of different dye solutions were measured using a colour flex spectrophotometer. At first, the maximum wavelength for each dye was detected by taking absorbance at a wavelength range of 400–700 nm. The wavelength at which a distinct peak was observed was considered as the maximum wavelength (λ max). Obviously, there should be different λ maximum for different dye sources. A wavelength of 410 nm was found to give the maximum optical density for *Madhuca latifolia* bark.

2.8 Optimization of Madhuca Latifolia Dye Extraction Time

To optimize the extraction time for *Madhuca latifolia* dye, the optical density was noted after 30, 45 and 60 min of boiling. The time at which optical density recorded maximum was selected.

2.9 Optimization of Mordanting Procedures

2.9.1 Optimization of Mordanting Method

Eri silk yarns were subjected to mordanting, viz. pre-mordanting, simultaneous and post-mordanting methods. Eri yarn was mordanted and then dyed by pre-mordanting method. An alkaline mordant solution was prepared by dissolving the required amount of mordant in water. The yarn was placed in this solution for 30 min at 60 °C, was made ready for dyeing.

Mordant and dye were applied simultaneously in the same bath in the simultaneous mordanting method. The eri silk yarn was dyed first followed by the mordanted in the post-mordanting method. Later the yarn was washed, rinsed and dried.

2.9.2 Optimization of Mordanting Time

Pre-mordanting method was selected to optimize the mordanting time. Eri silk yarns were mordanted at 60 °C for 30, 45 and 60 min. The mordanted samples were then dyed and the colour strength of these samples was recorded.

2.9.3 Optimization of Mordant Concentrations

To optimize the mordant concentration suitable for each dye, four concentration levels were taken into consideration. For alum 5, 10, 15 and 20%, for ferrous sulphate, stannous chloride and tartaric acid and chitin 1, 2, 3, 4 and 5% solutions were prepared separately. To optimize the concentration of mordants, pre-mordanting method was used. The colour strength values were checked in the colour flex spectrophotometer. Based on the colour strength and visual appearance of samples, two concentrations of each mordant were selected.

2.10   Optimization of Dyeing Methods

2.10.1   Dye Uptake

To determine the dye uptake of eri silk yarns, the colour strength (K/S) values were calculated using a computer colour matching system, i.e. Colour flex spectrophotometer [5].

2.10.2   Optimization of Dye Material Concentration

To optimize the concentration, separate containers with 200 ml of water in each were taken to maintain material to liquor ratio of 1:50. The dye material from 1 to 10% was weighed and placed in the container and boiled at 95 °C for the optimized time. Yarn weighing 4 g was then placed in the dye liquor and dyed for 30–45 min. The colour strength was recorded in a colour flex spectrophotometer for the dyed samples [2].

2.10.3   Optimization of Dyeing Time

The extracted dye liquors were taken in the M:L ratio of 1:50 in different containers. The K/S values of the dye liquor before dyeing were noted. The pre-mordanted or unmordanted yarns were immersed in separate dye liquors and dyed for 30, 45, 60 min, respectively. The samples were removed and the K/S values of the liquors after dyeing for 30, 45 and 60 min were noted [6].

2.10.4   After Treatment

The dyed samples were washed in a lukewarm detergent solution to remove the loose dye on the yarn and then rinsed thoroughly in water and dried.

2.10.5   Evaluation of Colour Fastness Tests

Colour fastness refers to the resistance of the colour textiles to different agents used in washing, sunlight, perspiration and rubbing to which the yarn or fabric is exposed to during manufacturing and subsequent use. It is important because it directly affects the serviceability of fabrics. The most serviceable conditions for which textile is generally exposed are sunlight, washing, crocking and perspiration [8].

Table 1 Optimization of method of extraction for *Madhuca latifolia* dye

S.No	Extraction method	K/S values at λ(max)
1	Aqueous method	9.4419
2	Alkaline method*	11.6964
3	Acidic method	9.1345

*Indicates the selected method

3 Results and Discussion

3.1 Extraction of Dye from Madhuca Latifolia Bark

The variables considered in dye extraction from a natural source included the method of extraction, extraction time, dye material concentration and dyeing time.

3.2 Optimization of the Method of Extraction for Madhuca Latifolia

The coloured pigment present in the bark of *Madhuca latifolia* was extracted by employing three common methods of extraction using water as a medium. Among these methods, the alkaline method yielded more pigment for this dye source than the other methods as evident from Table 1.

The colour strength (K/S) value (taken at suitable wavelength 410 nm) was found to be maximum for alkaline method followed by acidic and aqueous methods indicating the suitability of this method for *Madhuca latifolia* dye.

3.3 Optimization of Mordant Concentrations for Madhuca Latifolia

The colour strength of *Madhuca latifolia* dye material with varying concentrations from 1 to 10% was evaluated using a colour flex spectrophotometer, the values of which are presented in Table 2. For the selection of suitable dye concentration, both the colour strength and appearance were considered. Thus the concentration level of 4 g/100 ml dye was selected.

Table 2 Optimization of dye material concentration for *Madhuca latifolia*

S.No	Concentrations	K/S values at λ(max)
1	1.0	2.7216
2	2.0	3.1174
3	3.0	3.5349
4	4.0*	3.7246
5	5.0	2.5461
6	6.0	2.2925
7	7.0	3.3895
8	8.0	2.9099
9	9.0	3.0988
10	10.0	2.7921

* Indicates the selected material concentration

Table 3 Optimization of dye extraction time for *Madhuca latifolia* dye

S.No	Extraction time (min)	K/S values at λ(max)
1	45	3.7984
2	60*	4.0270
3	90	2.5557

*Indicates the selected dye extraction time

3.4 Optimization of Dye Extraction Time for Madhuca Latifolia

The data on the density of dye liquor by varying the time of extraction is presented in Table 3. It was evident that an increase in the time of extraction over 60 min had decreased the K/S value of the dye liquor. Hence, 60 min time was found appropriate for extraction of dye from the bark of *Madhuca latifolia* [10].

3.5 Optimization of Mordant Concentrations for Madhuca Latifolia

The data on optimization of mordant concentrations are presented in Table 4 for *Madhuca latifolia* dye.

For alum, concentration levels of 10 and 15% were selected. Similarly, concentration levels of 1% and 3% for ferrous sulphate, 3% and 4% for stannous chloride, 2% and 4% for tartaric acid were selected. In case of chitin, 4% and 5% concentrations were selected. Among the *Madhuca latifolia* dyed samples, 15% and 5% alum showed higher (7.8976) and lower (0.4441) K/S values, respectively. Pinkish

Table 4 Optimization of mordant concentrations for *Madhuca latifolia* dye

S.No	Name of the mordant	Mordant concentrations g/100 g eri silk yarn	K/S values at λ(max)
1	Alum	5	0.4441
		10*	2.9028
		15*	7.8976
		20	3.2542
2	Ferrous sulphate	1*	1.4631
		2	3.3585
		3*	3.7684
		4	4.0250
3	Stannous chloride	1	2.5067
		2	3.1851
		3*	2.4697
		4*	3.6912
4	Tartaric acid	1	2.3954
		2*	1.0464
		3	2.9023
		4*	1.6491
5	Chitin	2	1.9873
		3	2.5955
		4*	3.5466
		5*	6.5706

*Indicates selected mordant concentration

brown to dark brown colours were obtained by dyeing eri silk mordanted with various mordants using *Madhuca latifolia* dye.

3.6 Shade Variation in Madhuca Latifolia Dyed Eri Silk After Mordanting

Eco-friendly mordants such as alum, ferrous sulphate, stannous chloride, tartaric acid and chitin were used for mordanting eri silk yarn. The data on optimization of mordanting methods for all the mordants on eri silk yarn is represented in Table 5.

Bright brown shades were observed in alum pre-mordanted samples. Little light shades were produced in simultaneous and post-mordanted samples. Very dark brown shades were shown in ferrous sulphate mordanted samples. All three mordanting methods produced greyish brown shades, probably due to saddening effects produced with ferrous sulphate.

Table 5 Optimization of mordanting methods for all the mordants on eri silk yarns for *Madhuca latifolia* dye

S.No	Mordanting method	Mordant used	Mordant concentration gr/100 g of Eri silk yarn	K/S value at max
1	Pre mordanting	Alum	10	0.1985
			15	0.3788
		Ferrous sulphate	1	0.1688
			3	0.3077
		Stannous chloride	3	0.1238
			4	0.1765
		Tartaric acid	2	0.3719
			4	0.2194
		Chitin	4	0.3025
			5	0.3241
2	Simultaneous mordanting	Alum	10	0.295
			15	0.3012
		Ferrous sulphate	1	0.1309
			3	0.3034
		Stannous chloride	3	0.3569
			4	0.2751
		Tartaric acid	2	0.251
			4	0.3118
		Chitin	4	0.2878
			5	0.3189
3	Post mordanting	Alum	10	0.2986
			15	0.1478
		Ferrous sulphate	1	0.3528
			3	0.1683
		Stannous chloride	3	0.3804
			4	0.3271
		Tartaric acid	2	0.1702
			4	0.352
		Chitin	4	0.03092
			5	0.3257

Stannous chloride mordanted samples exhibited light brown shades in pre-mordanting method. Bright brown colours were observed in case of simultaneous method. Post-mordanted samples showed very dull brown shades.

Tartaric acid and chitin mordanted samples exhibited very good pinkish-brown to dark brown shades. All three mordanting methods produced very bright shades.

Table 6 Optimization of dyeing time for *Madhuca latifolia*	Sl.No	Extraction time (min)	K/S values at λ(max)
	1	45*	3.5213
	2	60	3.3997
	3	90	2.6428

*Indicates selected dyeing time

3.7 Optimization of Dyeing Time for Madhuca Latifolia

In case of *Madhuca latifolia*, 45 min dyeing time was found suitable as the highest K/S value was observed as compared with 60 and 90 min. So 45 min dyeing time was selected for *Madhuca latifolia* dye. The data is presented in Table 6.

4 Conclusion

Recently, eri culture was introduced in Andhra Pradesh due to the availability of host plants. Eri silk fabrics are suitable in the fields of apparel, furnishing and home textiles. Eco- friendly mordants such as alum, stannous chloride, ferrous sulphate, tartaric acid and chitin were selected for the study. The maximum wavelength of 410 nm was selected for *Madhuca latifolia* liquids. The optimum time for dyeing was 45 *min* for *Madhuca latifolia* dye source. Eri silk mordanted with various mordants and *Madhuca latifolia* dyed samples exhibited colours ranging from pinkish-brown, light brown to dark brown colours.

5 Implications of the Study

Nowadays availability and use of natural dyes in the present situation raises very big concerns about the sustainability of the perception. Less expensive production methods of natural dyes and reasonable industrial application are required. Huge water consumption and the amount of often unnecessarily lengthy working operations should be reviewed in the situation of cost savings and considerate approach to the environment. The study on *Madhuca latifolia* as sources of natural dye for eri silk is useful for the dyeing industry owing to the increased awareness and demand for eco-friendly products. Another important advantage is its adaptability in small-scale units due to easy dyeing procedures without the use of high-tech machinery. The cost of dyeing can be reduced further by commercialization. Different colours developed from *Madhuca latifolia* add a range of colours to the existing spectrum of colours of natural dyes with eco-friendly mordants. Therefore, craftsmen and dyers can adopt this technology to increase the colour range and produce variety for consumers. The research study findings clearly demonstrate that the extraction of natural colourants from the bark of plants can also be a sustainable technique towards waste utilization.

6 Suggestions for Further Study

1. The study can be taken up on eri blended fabrics.
2. Combinations of mordants can also be tried.
3. Further study can be taken up on printing and painting with *Madhuca latifolia* extracts.

References

1. Ashis NB, Kotnala OP, Maulik SR (2018) Dyeing of Eri silk with natural dyes in presence of natural mordants. Indian J Tradit Knowl 17(2):396–399. http://nopr.niscair.res.in
2. Bhandari NL, Pokhrel B, Bhandari U, Bhattarai S, Devkota A, Bhandari G (2021) An overview of research on plant based natural dyes in Nepal: scope and challenges. J Agricult Nat Resour 3(2):45–66. https://doi.org/10.3126/janr.v3i2.32328
3. Bhandari N, Bist K, Ghimire J, Chaudhary S, Pandey D, Adhikari R (2020) Feasibility study of the euphorbia pulcherrima plant extract as natural dye with different mordants for fabric dyeing. J Inst Sci Technol 25(1):30–36. https://doi.org/10.3126/jist.v25i1.29421
4. Deepti P, Shahnaz J, Manisha G (2020) Functional properties of natural dyed textiles. Chem Technol Nat Synth Dyes Pigments Intech Open. https://doi.org/10.5772/intechopen.88933
5. Gitanjali B, Ava Rani P, Shankar HG (2018) Eco-friendly Dyeing of Mulberry silk Yarn with Bark of Artocarpus lacucha. Int J Curr Microbiol App Sci 7(09):552–562. https://doi.org/10.20546/ijcmas.2018.709.066
6. Khan AH, Jiang MT et al (2018) Dyeing of silk fabric with natural dye from camphor (Cinnamomum camphora) plant leaf extract. Color Technol 134: 266–270. https://doi.org/10.1111/cote.12338
7. Ksanbok M, Marvellous BL (2020) Plant species used for dyeing Eri-silk and their conservation by the Bhoi Women of Meghalaya, India. Int J Ecol Environ Sci 46
8. Nabila T, Mohd RA, Muhammad IAK, Khudzir I (2019) Effect of mordant types and methods on the colour fastness properties of silk fabrics dyed with brown seaweeds. Int J Recent Technol Eng (IJRTE) 8(4). ISSN: 2277-3878. https://doi.org/10.35940/ijrte.D5173.118419
9. Samanta P (2020) A review on application of natural dyes on textile fabrics and its revival strategy. Chem Technol Nat Synth Dyes Pigments Intech Open. https://doi.org/10.5772/intechopen.90038
10. Saradi JG, Nabaneeta G, Binita BK (2019) Tea leaves: a herbaldye on ERI silk. J Pharmacogn Phytochem 8(6):442–446

Development of Mosquito Repellent Finish for Textiles Using Neem Oil: An Eco-Friendly Approach

Deepali Rastogi, Archana Jain, and Anupriya Negi

Abstract Finishing is an integral part of textile processing which augments the performance of the fabrics and adds functional properties as well. According to AMCA (The American Mosquito Control Association), mosquitoes are responsible for more than one million deaths per year worldwide. Many studies have been undertaken to develop mosquito repellent textiles using chemical and natural sources. Due to the negative effects of synthetic mosquito repellents, people are switching to herbal and natural plant-based mosquito repellents. The present study is an endeavour in this direction and is aimed at developing a mosquito repellent finish for textiles using neem oil extract. The formulation was developed in different concentrations for the optimization of the recipe. The shelf life of the formulation was assessed. The developed formulation was applied to different types of cotton fabrics using two methods of application. The efficacy of formulation was evaluated by conducting cage tests for the finished fabrics at NIMR (National Institute of Malaria Research, Delhi). The effect of storage time period of the finished fabric on mosquito repellency was also studied. The neem oil formulation gave promising results as it was effective against mosquitoes and could be applied as a household finish to various textiles.

Keywords Eco-friendly · Neem · Mosquito repellence · Textile application · Renewable finish

1 Introduction

The textile finishing industry is accountable for the consumption and discharge of huge quantities of chemicals and substances that are perilous to our environment. Furthermore, finishing is liable for creating the highest volumes of wastewater amidst

D. Rastogi · A. Negi
Department of Fabric and Apparel Science, Lady Irwin College, New Delhi, India
e-mail: deepali.rastogi@lic.du.ac.in

A. Jain (✉)
Vivekananda College, University of Delhi, New Delhi, India
e-mail: archanajainfas@gmail.com

© The Author(s), under exclusive license to Springer Nature Singapore Pte Ltd. 2022   183
S. S. Muthu (ed.), *Sustainable Approaches in Textiles and Fashion*,
Sustainable Textiles: Production, Processing, Manufacturing & Chemistry,
https://doi.org/10.1007/978-981-19-0538-4_9

all other stages of textile processing. Consumers in present times have inexhaustible expectations from textile products with sustainability as a prerequisite. There is a need to shift to practices that are eco-friendly and use raw materials that are non-hazardous to the environment.

Protective textile is one of the upcoming areas of technical textiles with mosquito repellence as one of its most desirable attributes. There are various ways in which mosquito repellents are used in the form of mats, lotions, creams, coils, patches, etc. Incorporating mosquito repellent formulation into the fabric is one of the revolutionary steps taken in the advancement of the textile finishing industry. The textiles with mosquito-repelling features can drive the mosquito away from the treated textile, or knock down the mosquitoes when it is in contact with the treated textiles. The addition of repellent properties to textile material can help to reduce the statistics of vector disease cases without having any side effects on the wearer.

It has been found that the use of synthetic repellents have various negative effects on the health of the people as they cause headache, dizziness, sore throat, nausea, vomiting, stomach pain and drowsiness, and can sometimes lead to asphyxiation and suffocation. DEET (*N, N-diethyl-m-toluamide*) and permethrin are the predominant synthetic mosquito repellents and permethrin has been used for finishing textiles since the advent of mosquito repellent textiles. Many studies have been undertaken to make advancements in this area since then. Permethrin is a pyrethroid-based insecticide similar to pyrethrins, a natural insecticide extracted from chrysanthemum flowers. Military uniforms for the soldiers working in areas where they are vulnerable to attack by various insects are constructed with permethrin incorporated textiles. Studies have been conducted to improve the longevity of finish efficacy on cotton textiles by employing techniques like micro- and nano-encapsulation of limonene and permethrin [1].

As DEET applied to textiles does not have wash fastness even for one cycle, trials have been made to synthesize its derivatives and coupling with different naphthols to get dyed cotton fabric, imparted mosquito repellent finish by combining the steps of dyeing and finishing into a single process [2]. The application of chemical-based repellents for prevention from insects and arthropods has given rise to various concerns as they are not eco-friendly and are precarious for human beings, animals and aquatic species [3]. Due to the negative effects of synthetic mosquito repellents, people are switching to more herbal and natural plant-based mosquito repellents. The easy availability of these plant-based repellents has made these repellents preferential over synthetic ones.

There are many plants having mosquito repellent properties such as neem, basil, catnip, marigold, rosemary, clove, eucalyptus, citronella, sweet orange, etc. [4]. Many studies have been conducted to see the effectiveness of various plants as mosquito repellents. Development of mosquito repellent cotton fabric using eco-friendly mint [5], castor oil [6], marigold petals [7, 8], sweet basil and eucalyptus [9, 10], citronella [11] and different medicinal natural plants [12] was investigated and they were found to be effective as mosquito repellents though they were poor in wash fastness.

Researchers have been investigating different methods for the application of these plant and chemical-based finishes on textiles to boost their performance.

Four methods that are employed to impart these finishes include absorption (padding/dipping/spraying), incorporation, polymer coating and microencapsulation [13]. Microencapsulated citronella oil [14] and other herbal oils like lemongrass [15] were found to have improved mosquito repellence longevity.

The innumerable researches administered to utilize the non-hazardous herbal plant sources clearly stipulate their applicability as mosquito repellent finishes for textiles. Though there are some limitations like the durability of finishes, they could be condoned owing to our concerns for sustainability. Efforts should be made to modify the techniques used to impart finishes to the textiles for enhancing their durability.

In this study, an attempt was made to develop a formulation based on neem oil which can impart mosquito repellence to textiles and can be used both at commercial and household levels. This formulation can be used by people who are vulnerable to mosquito bites like people working in forest areas and children who usually spend the major part of their day outside. Application of such formulation on apparel and home textiles makes it easy and effective without having a need to carry mosquito repellents separately in the forms like creams, sprays, lotions, etc. In addition, it can be used after laundering every time like other blueing or starching agents which will overcome the problem of lack of durability, the most obvious concern with respect to herbal mosquito repellent finishes.

This chapter gives an overview of the methodology followed for developing mosquito repellent formulations from Neem oil and evaluation of the shelf life of the formulation. This chapter includes the details of the application of the developed formulation to the fabric samples and the evaluation of the samples. The pad-dry and spraying techniques used for the application of the formulation to the fabrics have been discussed. Performance of the finished fabrics with respect to various factors like mosquito repellency, toxicity, mosquito repellency after storage and after laundering as well as the change in colour of the fabric after the application of finish have been discussed.

2 Methodology

This study involved the preparation of neem oil formulation for finishing textiles to obtain mosquito repellence properties. The finished fabric was evaluated for its mosquito repellence properties using the cage test. The effect of laundering on the durability of formulation on finished fabric was studied. The shelf life of the formulation was also assessed. The research design of the study comprised of two phases:

Table 1 Basic composition of Neem oil mosquito repellent formulation

S. No	Ingredients	Category	Quantity
1	Neem oil	Essential oil	20–40 ml
2	Sodium Lauryl Ether Sulphate	Emulsifier	10–40 ml
3	Sodium Benzoate	Preservative	0–0.5 g (with/without)
4	Water		To make it 100 ml

2.1 Phase I: Preparation of Mosquito Repellent (MR) Formulation and Assessment of Its Shelf Life

The formulation based on neem oil was prepared in various compositions and the shelf life of the final formulation was assessed with and without stabilizers.

A. **Preparation of the formulation**. The first and foremost step was to prepare the formulation of neem oil using an emulsifier. The composition of the formulation is given in Table 1

B. **Assessment of shelf life of Mosquito Repellent formulation**. Shelf life is a period during which a good remains effective and free from deterioration, and thus saleable in the given standard conditions. The shelf life of the formulation was evaluated by observing it for any signs of deterioration or fungal growth for a period of time.

   The assessment was carried out by keeping the formulation at room temperature for 1–12 weeks. Also, the effect of sodium benzoate as a preservative on the mosquito repellent formulation was studied. The formulation was divided into two parts. In one part sodium benzoate was added and the other part did not contain any preservative. The shelf life of both the formulations was studied for 12 weeks.

2.2 Phase II: Application and Evaluation of Mosquito Repellent Formulation

This phase comprised the application of the Neem Oil formulation to two types of fabrics and the evaluation of their efficacy against mosquitoes. Various physical parameters of the finished fabrics were also evaluated.

A. **Application of formulation**: The formulation prepared was applied to two different fabrics, namely cambric and casement, using two different techniques, i.e. Pad-Dry technique (2 dips-2 nips) and Spray technique. The formulation was applied in three different concentrations (10, 20 and 30%) by mixing with the required quantities of water on the selected fabrics, i.e. cambric and casement.

Fig. 1 Set up for cage test

B. **Evaluation of the finished fabrics**. The following test was performed on the finished fabrics.

Assessment of mosquito repellence activity: The mosquito repellency properties were studied using a standard cage test which was performed at the National Institute of Malaria Research, Dwarka, New Delhi.

- **Cage test** is also known as Arm-in-cage test, which is a standard test method for investigating the efficacy of mosquito repellent formulations. Cage tests are quick and cost-effective ways to determine the mosquito-repelling qualities of treated textiles. The repellence test of the samples was done on the basis of the standard as certified by WHO with some modifications that were thought to be important for the study.

    The repellence test was done by preparing a cage (Fig. 1) and rearing 30 mosquitoes ((laboratory reared *Anopheles stephensi*, 3 days old). The 40 cm × 40 cm ferrous frame cage was covered with muslin cloth and its front was covered with a nylon net. The samples were inserted into the cage without inserting the forearms of the volunteer and cotton soaked in glucose was placed in the cage to confuse mosquitoes of the human blood smell. While the mosquitoes were inserted into the cage with a suction tube, they tend to sit/rest on the wall of the cage and not on the ground where the samples were kept. So, the testing method was modified by placing the treated samples on the wall of the cage and the number of mosquitoes that arrived on the treated samples was counted and recorded for four hours. The cage was covered with a black colour cloth ensuring that there is no light entering the cage. Also, mosquitoes settle on a place where they sit first (when it is convenient for them), they may not fly to the other place. So, the cage was shaken each 10 min to disturb the mosquitoes. The number of mosquitoes found sitting on the fabric after every hour were counted. The observation was done for up to four hours.

Table 2 Conditions for washing treated fabric samples

Amount of water (ml)	Amount of soap (g/l)	Time (min)	Type of wash
200	5	30	Cold wash

Protection % was calculated as follows:

$$\text{Protection \%} = \frac{\text{No. of mosquitoes on control sample-no of mosquitoes on treated sample}}{\text{No. of mosquitoes on control sample}} \times 100$$

- **Effect of storage time on mosquito repellence of treated fabrics**: Effect of storage time on mosquito repellent efficiency of the treated fabrics was studied using the cage test. The effectiveness of the formulations on the fabrics was evaluated within 24 h and after 15 days.
- **Effect of laundering on mosquito repellence of treated fabrics**: The treated fabric samples were washed in a launderometer according to the conditions given in Table 2. After washing, the fabric samples were tested again for their mosquito repellence properties using the cage test to check whether the finish is durable or not.
- **Cone bioassays test**-For this, the rim of the cone was fixed to the treated fabric samples fastened with a rubber band. The cones were fixed randomly with fabrics treated with different concentrations of finish. One fabric sample from each concentration and one control sample was taken. Ten mosquitoes were introduced into each of the plastic cones through an orifice and the orifice was plugged with a cotton ball. Two exposures were made for three minutes on two flaps of the treated fabric and the control. The number of mosquitoes knocked down at the end of the three minutes were recorded and the mosquitoes were transferred to plastic containers with a nylon net fastened with a rubber band. Mosquitoes were provided with the sugar-soaked cotton ball placed on the top of the net. The plastic container was placed preferably in an unsprayed room maintained at standard temperature (27 ± 2 °C) and relative humidity (60–70%). If it is not feasible to maintain temperature and humidity, a moist chamber can be used as described in WHO (1981). Mosquitoes are considered to be alive if they can both stand upright and fly in a coordinated manner. Mosquitoes that are moribund or dead are classified and recorded as knocked down at 60 min and as dead at 24 h. A mosquito is moribund if it cannot stand (e.g. has one or two legs), cannot fly in a coordinated manner or takes off briefly but falls immediately. A mosquito is dead if it is immobile, cannot stand or shows no sign of life. Per cent mortalities are determined after 24 h of holding from the alive and dead mosquitoes. Results are expressed as overall persistence against the given dose of insecticide.

$$\%\text{Mosquito Mortality} = \frac{\text{No. of mosquitoes knocked dead}}{\text{Total no. of mosquitoes introduced}} \times 100$$

3   Results and Discussion

3.1   Preparation of Formulations and Evaluation of Their Shelf Life

The formulation was prepared using four ingredients, viz. Sodium Lauryl Ether Sulphate, Neem oil, Sodium Benzoate and water. Sodium lauryl ether sulphate works as an emulsifier which helps in making an emulsion of essential oil and water, whereas sodium benzoate acts as a preservative. To check the shelf life, neem oil formulation was divided into two parts. In one formulation, sodium benzoate was added and the other was left as it is for the time period of a minimum of four months. There was no sign of fungal growth or deterioration in both cases (with and without sodium benzoate) even after 4 months of storage at room temperature. Hence. this formulation can be used for apparel and upholstery in houses and can be kept for long periods. The only precaution which should be kept in mind is that the shaking of the formulation is important before use.

3.2   Evaluation of the Finished Fabric Samples

Mosquito repellency of the finished fabrics was determined at the National Institute of Malaria Research, Dwarka, New Delhi, using Cage test methods. The results are discussed below.

Cage test. The test was done for a duration of 4 h. The number of mosquitoes landing on the test samples was counted after every hour for up to four hours and the total number of mosquitoes was then calculated. The effectiveness of the sample was evaluated on the basis of the number of mosquitoes attracted on the finished and control fabric samples The test was done on the Casement and Cambric samples finished with different application methods-Pad-Dry method and Spraying method. The finished fabrics were tested within 24 h of the treatment and after 15 days of the treatment to study the change in efficacy on storage, if any.

Results for samples tested within 24 h of application of finish: The mosquito repellency results of cage test (within 24 h) of the Cambric and Casement fabric treated with neem oil formulation by padding method are given in Tables 3 and 4.

Table 3 Mosquito repellency of Cambric fabric finished using pad-dry technique (within 24 h)

Concentration of finish	Number of replicates	Number of mosquitoes landing (after 1–4 h)						Protection percentage (%)
		1	2	3	4	Total	Mean	
Control	i	7	5	5	9	26	25.5	
	ii	4	6	7	8	25		
10%	i	2	2	3	4	11	10.5	58.82
	ii	3	1	2	4	10		
20%	i	0	1	2	3	6	7	72.54
	ii	1	2	2	3	8		
30%	i	0	0	1	2	3	4	84.31
	ii	1	2	0	2	5		

Table 4 Mosquito repellency of casement fabric finished using pad-dry technique (within 24 h)

Concentration of finish	Number of replicates	Number of mosquitoes landing (after 1–4 h)						Protection percentage (%)
		1	2	3	4	Total	Mean	
Control	i	5	7	6	9	23	22	
	ii	4	6	4	7	21		
20%	i	1	2	2	3	7	7	68.18
	ii	0	1	2	4	7 ara>		
30%	i	0	1	1	4	4	4	81.81
	ii	0	0	2	2	4		

The finished cambric fabric showed a significant increase in mosquito repellency with an increase in the concentration of the finish. On application of 10% concentration of formulation, approximately 60% protection was achieved which significantly increased to 72% on increasing the application concentration to 20%. With 30% concentration, almost 84% of mosquito repellency could be achieved.

10% concentration was not used for further experiments as it showed very poor results.

As can be seen from Table 4, the mosquito repellency seen on treated casement fabric is slightly less than that on cambric fabric, however, the difference is not very significant.

Tables 5 and 6 show the results of samples which were treated by spray method and were tested within the time period of 24 h.

The results on both the fabrics show that there is an increase in protection percentage as the concentration of oil increases but the protection achieved by spraying method is much lower than that obtained by padding technique. This could be due to lesser amount of finish being received by the fabric as a significant amount

Table 5 Mosquito repellency of Cambric fabric treated using spray technique (within 24 h)

Concentration of finish	Number of replicates	Number of mosquitoes landing (after 1–4 h)						Protection percentage (%)
		1	2	3	4	Total	Mean	
Control	i	6	5	6	7	24	24	
	ii	5	4	7	8	24		
20%	i	2	3	2	4	11	9.5	60.41
	ii	1	3	2	2	8		
30%	i	3	2	1	2	8	8	66.66
	ii	1	1	3	3	8		

Table 6 Mosquito repellence of casement fabric finished with spray technique (within 24 h)

Concentration of finish	Number of replicates	Number of mosquitoes landing (after 1–4 h)						Protection percentage
		0–1	1–2	2–3	3–4	Total	Mean	
Control	i	6	5	7	8	26	25	
	ii	6	4	8	6	24		
20%	i	2	3	3	4	12	9	64.0
	ii	2	1	2	1	6		
30%	i	0	1	2	4	7	7.5	70.0
	ii	0	2	3	3	8		

of the finish may get lost in the air during spraying. In spite of this limitation, 65–70% protection could be achieved when 30% formulation was applied. The casement fabric gave better results for mosquito repellence than the cambric fabric when the finish was applied by spraying technique.

This technique can be very beneficial when one wants to impart mosquito repellency to textile products, which are difficult to treat by pad-dry method, such as upholsteries, curtains, carpets, mattresses, etc.

Effect of storage time on mosquito repellence properties of the treated fabric. A cage test was used to see the efficacy of the treated fabrics after 15 days of storage which gave an idea about the efficacy of formulation on the fabrics on storage. The results of cambric and casement fabrics treated with mosquito repellent formulation and tested after 15 days of application are given in Table 7.

In both the fabrics, there was a considerable reduction in the mosquito repellency, the protection percentage reduced to around 60% for fabrics treated with 20% concentration and to 70% in the case of 30% formulation.

Table 7 Mosquito repellency after 15 days of treatment by padding method

Concentration of finish (%)	Protection Percentage (%)			
	Cambric		Casement	
	Within 24 h	After 15 days	Within 24 h	After 15 days
20	72.54	60.37	68.18	60.7
30	84.31	71.69	81.81	70.5

Table 8 Effect of laundering on mosquito repellency of the treated fabric

Concentration of finish (%)	Protection percentage (%)			
	Cambric		Casement	
	Before wash	After wash	Before wash	After wash
20	72.54	41.66	68.18	55.76
30	84.31	47.91	81.81	53.84

3.3   Effect of Laundering on Mosquito Repellency of Treated Fabric

The efficacy of mosquito repellent formulation was determined after washing the treated fabrics. For this, the treated fabric samples were washed and tested again to check whether the finish is durable (Table 8).

It was observed that after washing, the protection of treated fabric reduced significantly in both cambric and casement. However, cambric fabric showed a greater reduction in the protection percentage than the casement fabric.

Overall, casement fabric, among the two fabrics used for the study, gave better results, which means it showed a better protection percentage than cambric fabric at all the concentrations.

3.4   Cone Bioassays Test

To evaluate the toxicity of the finished fabric, the cone bioassays test was used. The toxicity of the formulation was evaluated on the basis of mortality reported after 24 h of feeding glucose to the knock down mosquitoes (Table 9).

It was observed that the treated fabrics were not toxic to mosquitoes. The formulation applied on the fabrics only repels the mosquitoes through unwanted smell without killing them.

Table 9 Cone bioassays test on fabric treated with 20 and 30% formulation

Fabric	No. of replicates	No. of mosquitoes expressed in each cone	Knock down in 3 min expression		No. of mosquitoes knocked dead after 24 h	
			20%	30%	20%	30%
Cambric	i	10	1	1	0	0
	ii	10	0	1	0	0
Casement	i	10	0	0	0	0
	ii	10	1	2	0	0

3.5   Effect of Formulation on the Colour of the Treated Fabrics

The change in colour due to the application of the finish was studied by testing the fabrics on a spectrophotometer. The results were assessed on the basis of K/S, L* C* h* values, Whiteness Index and Yellowness Index (Table 10).

The K/S values obtained for all the treated samples were higher as compared to the control samples irrespective of the fabric type, concentration of formulation and method of application of finish as shown in the results above (Table 10).

The whiteness index and the yellowness index also indicate that there is yellowing of all the treated samples. The whiteness index of the treated samples is lower as compared to that of the control sample and an increase in the concentration of formulation resulted in further lowering of the whiteness index. The yellowness index of treated fabrics is higher as compared to control samples for all the variations. The fabrics given spraying treatment showed more yellowness.

Table 10 Colour assessment of the fabrics finished with neem oil formulation

Sample	Method of application	Concn	K/S	L*	C*	h*	WI	YI
Casement	Control	–	0.63	91.83	6.87	90.27	58.09	12.82
	Pad-dry	20%	1.29	89.24	8.21	90.30	41.02	20.95
		30%	1.67	87.91	8.31	88.87	34.28	24.02
	Spraying	20%	1.29	88.69	7.60	88.63	40.95	20.08
		30%	1.67	88.17	7.39	87.17	39.55	20.47
Cambric	Control	–	0.79	92.32	0.94	300.0	89.81	−1.54
	Pad-dry	20%	0.92	90.00	2.77	93.15	72.25	5.25
		30%	1.06	89.17	3.40	90.37	66.70	7.13
	Spraying	20%	1.25	88.72	4.74	95.44	61.49	10.05
		30%	1.78	87.47	6.95	95.43	47.64	17.08

4 Conclusion

The results of the study show that neem oil formulation can be used as a mosquito repellent finish for textiles effectively. This formulation being an alternative to synthetic mosquito repellents like DEET, permethrin, etc. will reduce the environmental impact of harmful chemicals which are hazardous for health and the environment. The only limitation it poses is the slight yellowing of white fabrics which makes this finish better suited for coloured fabrics. It can be used easily as a household finish for textiles like upholstery or home furnishings, especially at public places (hospitals, etc.). The main strength of this finish would be its ease of application by common consumers on a variety of fabrics as per the requirement.

References

1. Türkoğlu GC, Sarıışık AM, Erkan G, Yıkılmaz MS, Kontart O (2020) Micro- and nano-encapsulation of limonene and permethrin for mosquito repellent finishing of cotton textiles. Iran Polym J 29(4):321–329. https://doi.org/10.1007/s13726-020-00799-4
2. Teli MD, Chavan PP (2017) Dyeing of cotton fabric for improved mosquito repellency. J Text Inst 109(4):427–434. https://doi.org/10.1080/00405000.2017.1351066
3. Raja A, Kawlekar S, Saxena S, Arputharaj A, Patil PG (2015) Mosquito protective textiles—A review. Int J Mosq Res 2:49–53
4. Agarwal A (2015) Development of mosquito repellent wet wipes. Unpublished Master Dissertation, Department of Fabric and Apparel Science, Lady Irwin College, University of Delhi
5. Gupta A, Singh A (2017) Development of mosquito repellent finished cotton fabric using eco friendly mint. Int J Home Sci 3:155–157
6. Tseghai GB (2016) Mosquito repellent finish of cotton fabric by extracting castor oil. Int J Sci Eng Res 7:873–878
7. Rastogi A (2014) Eco-friendly Dyeing of Juco fabrics using waste Marigold (*Tagetes Erecta*) flowers for development of home textiles. Unpublished Master Dissertation, Department of Fabric and Apparel Science, Lady Irwin College, University of Delhi
8. Gupta A, Singh A (2016) Development of eco friendly mosquito repellent cotton fabric. Int J Curr Res 9(07):53434–53435
9. Kantheti P, Rajitha I, Padma A (2020) Natural finishes on textiles to combat the mosquitoes: a pilot study. J Entomol Zool Stud 8:30–33
10. Gogia M (2013) Application of Dye extracted from Eucalyptus *Citridora* leaves on selected synthetic fabrics. Unpublished Master Dissertation, Department of Fabric and Apparel Science, Lady Irwin College, University of Delhi
11. Kakaria R (2019) Mosquito repellent formulations from eucalyptus and citronella oil: a comparative study. Unpublished Master Dissertation, Department of Fabric and Apparel Science, Lady Irwin College, University of Delhi
12. Mia R, Sajib MI, Banna BU, Chaki R, Alam SS, Rasel A, Islam T (2020) Mosquito repellent finishes on textile fabrics (woven & knit) by using different medicinal natural plants
13. Anuar AA, Yusof N (2016) Methods of imparting mosquito repellent agents and the assessing mosquito repellency on textile. Fash Text 3(1). https://doi.org/10.1186/s40691-016-0064-y

14. Specos MM, García JJ, Tornesello J, Marino P, Vecchia MD, Tesoriero MV, Hermida LG (2010) Microencapsulated citronella oil for mosquito repellent finishing of cotton textiles. Trans R Soc Trop Med Hyg 104(10):653–658. https://doi.org/10.1016/j.trstmh.2010.06.004 PMID: 20673937
15. Thite AG, Gudiyawar DM (2015) Development of microencapsulated ecofriendly mosquito repellent cotton finished fabric by natural repellent oils. Int J Sci Technol Manag 4(11):166–174

Printed in the United States
by Baker & Taylor Publisher Services